Beyond the Gearbox: Vernier PM Machines Revolutionize Wind Turbine Design

Nicholas

First Printing, 2024

Table of Contents

1. Introduction

1.1 Background to the Study

The call from the wind turbine industry is for a machine which offers high torque at low speed, high torque density, high efficiency and reasonable material consumption. This book explores one such option which satisfies these criteria – the Vernier permanent magnet machine. Two classes of wind turbine schemes exist – fixed speed and variable speed turbines [1]. The Vernier topology falls into the latter category. The fixed-speed scheme, shown in Figure 1-1, incorporates a squirrel cage induction generator and a gearbox along with mechanical control techniques such as pitch and/or stall control to keep the rotor speed fixed at the generator rated speed. These generators are connected directly to the grid, and therefore the grid frequency determines the generator speed.

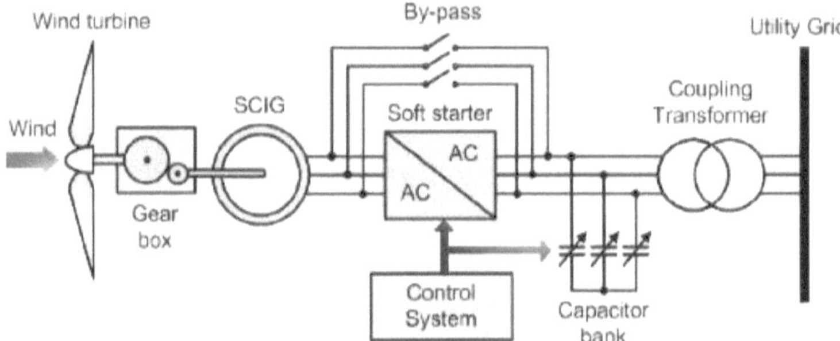

Figure 1-1: Fixed Speed Wind Turbine [2]

Structurally simple and robust, this scheme fell-short due to its suffering of heavy mechanical stress in attempting to maintain a fixed speed. A soft-starter is employed to temporarily reduce the torque in the power train during start-up by limiting the inrush current. The capacitor bank compensates for the reactive power drawn from the grid. More often than not, these induction generators have two windings, each with a different number of poles. This allows for operation at two different rated speeds.

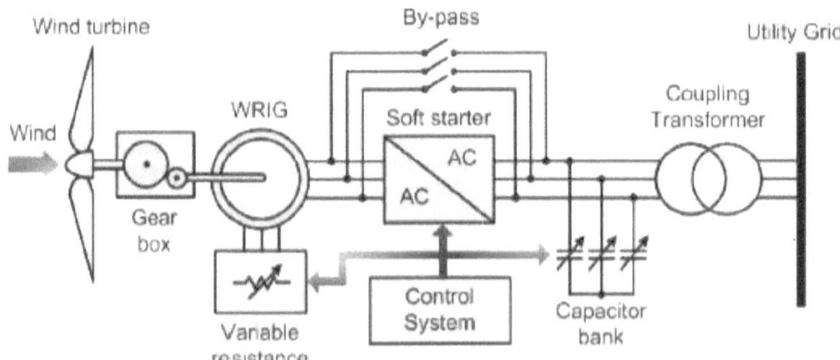

Figure 1-2: Partial Variable Speed Wind Turbine with a WRIG and variable rotor resistance [2]

The Wound Rotor Induction Generator (WRIG) case, or partial variable-speed topology, is shown in Figure 1-2. It is a variant of the fixed-speed turbine topology. This topology is analogous to the squirrel cage case in that the generator is connected directly to the grid. The only variation is the variable resistance bank which controls the rotor resistance. This is how speed control is achieved beyond rated speed. In fact, the size of the resistors determines the size of the variable speed range.

In fixed speed and partial variable speed schemes large gearboxes are necessitated by the matching of the low wind speed to the generator rated speed. However, some of the downsides of gearbox implementation include high maintenance and noise generation; which renders them operationally costly and repellent to nearby inhabitants [3]. As a result, direct-drive machines are preferred so as to alleviate these shortcomings. Unfortunately, to compensate for operating at a lower speed, direct-drive generators are designed to be much larger than their geared counterparts.

A decent compromise is the variable speed scheme which produces constant voltage at grid frequency under variable rotor speeds. This is accomplished using power converters [4]. Either induction or permanent magnet synchronous generators (PMSG) are used in this scheme. The variable speed scheme illustrated in Figure 1-3 below.

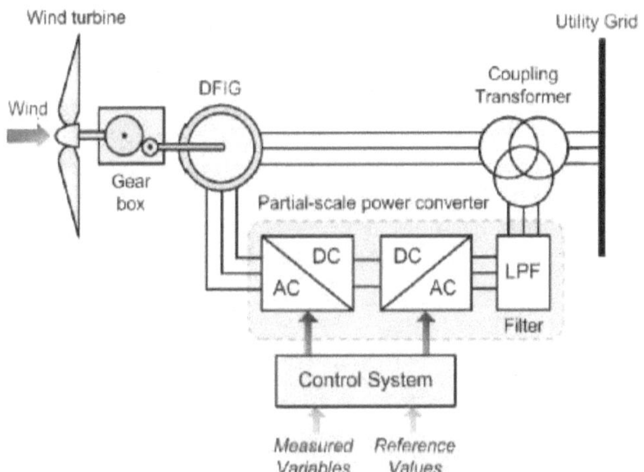

Figure 1-3: Variable Speed Wind Turbine with DFIG and Partial-Scale Power Converter [2]

If induction generators (either DFIG or WRIG) are used, the converter is connected to the rotor terminals, allowing for speed control via the slip. Either a full-scale or a partial-scale converter can be used. If a partial-scale converter is used, the converter is connected to the rotor windings of the WRIG.

With the advent of rare-earth permanent magnets [5], PMSGs became a viable option to be considered in wind turbine applications. PMSGs possess a substantially greater power density than induction generators. The implication being that PMSGs can be directly coupled to the turbine rotor blade hub without a gearbox. This is referred to as a direct-drive wind turbine configuration.

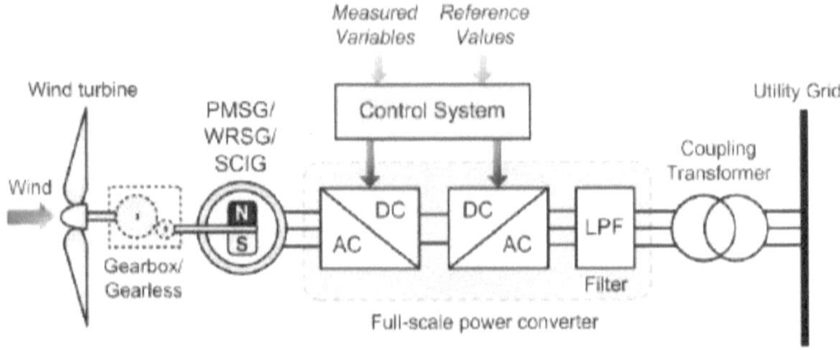

Figure 1-4: Direct-in-Line Variable Speed Wind Turbine with a Full-Scale power converter [2]

A full-scale power converter is used in conjunction with the synchronous generator in the direct-drive configuration as seen in Figure 1-4. The converter is connected in-line with the turbine transformer. Its role is to convert the variable frequency AC into fixed frequency AC, and also provide reactive power compensation.

This superior power density of the PMSG system is attractive to turbine designers. However, all turbines have one unescapable flaw: wind turbines are inherently inefficient. It will be shown here briefly that the high power density of PMSG systems does little to compensate for the turbine inherent inefficiency. This limitation is known as the Betz Limit. The Betz limit is an indication of the maximum power that the turbine can extract from the wind irrespective of the turbine design. It states that a wind turbine cannot extract more than 16/27 (59.3%) of the winds energy [6]. Consider the output power of a wind turbine [2]:

$$P = \frac{1}{2} C_p \rho A v^3 \tag{1.1.1}$$

C_p is the coefficient of performance, ρ is the air density, A is the swept area of the turbine, and v is the wind speed. The coefficient of performance represents the fraction of power extracted from the wind by the turbine. It is not constant and is dependent on wind speed, rotor speed, and pitch angle. Turbines use pitch control to hold C_p at its largest possible value to maximise output up to rated speed. Once rated speed is reached, C_p is reduced to maintain desired output power and prevent mechanical damage at high wind speeds. Rotor speed and wind speed are combined into a single variable to qualitatively analyse the performance coefficient C_p. This variable is termed the tip speed ratio:

$$\lambda = \frac{r_m \omega_m}{v} \tag{1.1.2}$$

r_m is the radius of the turbine blades. The tip speed ratio is controlled via the rotor speed to ensure that the turbine operates at maximum power coefficient. This is accomplished by Maximum Power Point Tracking (MPPT) techniques [7]. It is ever the goal of the wind turbine designer to push the limitations of the Betz Limit by designing generators with appreciable power density. This book will explore how well the Vernier topology, a member of the PMSG family, tackles this goal and how it is suited for a wind turbine application.

1.2 Objectives of this Study

1.2.1 Problems to be Investigated

In addition to power density, there are other determining factors in wind turbine suitability. One of which is cogging torque. Cogging torque is unique to PMSG machines. Cogging torque is due to the magnetic attraction between permanent magnets and the stator teeth. It has the effect of increasing the turbine cut-in speed thereby decreasing the wind-speed range in which power is produced [8]. Moreover, cogging torque exacerbates torque ripple during operation resulting in further mechanical loss, vibrations and noise. Additionally, voltage regulation is of particular importance in generator mode. Voltage regulation is the percentage difference in no-load back-EMF (Electro-Motive Force) and terminal voltage. It is a measure of how well the generator responds to being coupled to a load [9]. The underlying factor is the power factor which, when low, produces a high voltage regulation and requires a converter with a larger kVA rating [10]. Considering the aforementioned performance factors, a PMSG for a low-speed direct-drive wind turbine application has the following requirements:

- High power/torque density;
- High efficiency;
- Low cogging and ripple torque;
- Adequate power factor;
- High reliability, simple mechanical structure, low maintenance.

1.2.2 Purpose of the Study

In the world of direct-drive machines, the Vernier machine is an attractive option for its inherently high power density, capable of delivering high torque at low speeds [11] – the basic requirement of a direct drive machine. The Vernier topology has garnered interest in the electric vehicle [12] and wind turbine fields [13]. Operating under the "magnetic gear principle" (similarly to magnetic gears and magnetic geared integrated machines), the Vernier machine uses an additional air-gap flux density harmonic, the slot harmonic (or so-called "Vernier Component"), to supplement the fundamental torque and back-EMF. To accomplish this "magnetic gearing" effect the machine is wound with a different number of winding pole-pairs (p_s) to rotor pole-pairs (p_r), and the stator slots (Z) are selected such that:

$$p_r = Z - p_s \tag{1.2.1}$$

The result is a greater power density than a conventional PM machine [14]. Vernier machines can produce torque ripple as low as 0.2% without the use of torque ripple reducing techniques such as skewing [15]. Furthermore, the induced voltage of the Vernier machine, is inherently sinusoidal, achieving this without chording [16]. The magnetic gear effect has attracted significant attention. The principle has been implemented in the form of a stand-alone system as a replacement for mechanical gears [17] and as an addition to PM machines in the form of an integrated system termed a magnetic gear integrated machine (MGIM) [18]. Both innovations are practically ideal for wind power applications due to their contact-less speed changing mechanism. Structurally, the Vernier machine is significantly simpler than its MGIM counterpart. The MGIM topology usually consists of two rotating parts and two air-gaps; sometimes even three [19]. An example topology of the magnetic geared integrated machine is shown in Figure 1-5.

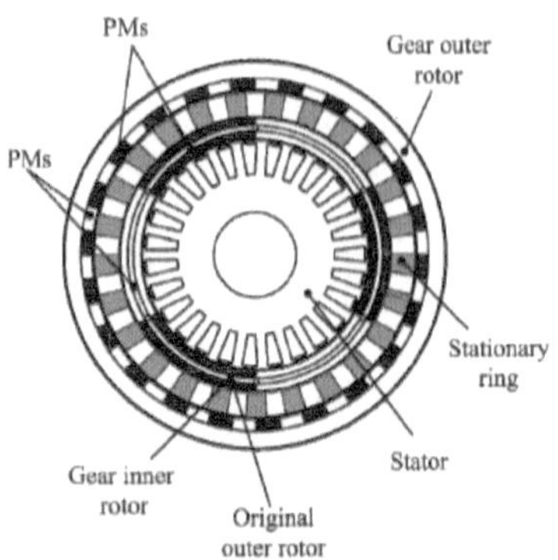

Figure 1-5: Magnetic Geared Integrated Machine structure [18]

On the other hand, the Vernier machine may consist of a single rotor and one air-gap. Hence the Vernier machine combines the high torque density of magnetic geared integrated machines with the structural simplicity of conventional surface-mounted machines [20]. Yet, there is one undesirable characteristic of Vernier machines – low power factor [21]. An unavoidable consequence of the Vernier superior power density, much research has been done to determine the root of this issue and design a remedy. As of late, research has shown that the double-stator Vernier machine topology alleviates the low power factor issue by reducing the harmonic leakage and thus also increasing power density [22] [23] [24]. The double-stator topology increases power density by 50% than that of single-rotor Vernier machines. The topology is shown in Figure 1-6. This is accomplished by employing two stators, one on either side of the rotor. This setup reinforces the magnetic circuit thereby reducing flux leakage, increasing power density and improving power factor. Later on, it will be shown that there even exists a trade-off between power factor and power density. A wind turbine designer using a Vernier PM generator must choose between power density and power factor bearing in mind that a low power factor requires the coupling of a larger (and thus costlier) power converter to the machine terminals.

Figure 1-6: Double Stator VPM structure [22]

To conclude, the Vernier Machine offers a means to eliminate the need for mechanical gears in wind turbine systems; thereby reducing maintenance requirements, noise and generally improving reliability. To this end, the Vernier machine is an attractive option in direct-drive applications as it has the potential to provide high power density and low torque ripple at low speeds. The lone drawback is the poor factor, this paper will evaluate whether the Vernier topology's other benefits will outweigh this solitary setback.

1.3 Scope and Limitations

This book constitutes a comparative study of four Vernier PM machine topologies. Based on this study, the most suitable topology will be chosen for prototyping. As such this book does not include any optimization studies. Finite Element Analysis (FEA) simulations are used extensively in these studies. FEA studies are notoriously time and computing power intensive. Rather, the prototype is designed based on what is practically realisable. The practical realisation of the prototype is dependent on cost and manufacturing limitations.

1.4 Plan of Development

To explore this Vernier topology and its role in wind energy, a comparative study is conducted using two different rotor types and two different stator types, namely; spoke-type and surface-mounted and fractional slot and integral slot respectively. Vernier theory is used to size the designs after which 2D FEM simulations are used to analyse and validate the designs. The main outcome of the comparative study is an assessment of the suitability of the two topologies for the direct-drive wind application. The design will be assessed on parameters which are critical for wind turbine functionality, namely; efficiency, torque ripple, torque density and material utilisation. All-in-all, four 6kW 250RPM machines are analysed and compared:

1. Integral-slot surface mounted (ISSM) machine with 12 slots and 22 poles
2. Integral-slot spoke type (ISST) machine with 12 slots and 22 poles
3. Fractional-slot surface mounted (FSSM) machine with 18 slots and 28 poles
4. Fractional-slot spoke-type (FSST) machine with 18 slots and 28 poles

The most suitable topology is chosen for prototyping. The essence of power factor and behaviour with respect to torque is studied in further detail. After which, a practically realisable design is put forward for the prototype and re-analysed using FEA. The prototype is manufactured and tested. The prototype is subsequently tested in the UCT machines lab and the results are analysed and compared to the analytical and FEA calculations.

2. Literature Review

The literature review begins with a history of the Vernier topology. It then goes on to critique the current state of Vernier machine design theory. It will be evident that Vernier machine design theory is not as well defined as conventional PM machine theory.

2.1 Vernier Reluctance Motor – Humble Beginnings

The Vernier motor topology has its roots in the doubly salient reluctance machines. The earlier Vernier machines were designed using the same structural blueprint as the doubly salient machines. Such machines were termed Vernier reluctance machines [VRM]. The Vernier reluctance motor operates on the principle of a doubly salient reluctance machine with different number of teeth on the stator and rotor. The difference in number of teeth prevents the rotor from aligning with the stator in static equilibrium. The patent [25] constitutes the design of such a machine with 48 stator teeth and 50 rotor teeth. This pioneering Vernier topology is shown in Figure 2-1.

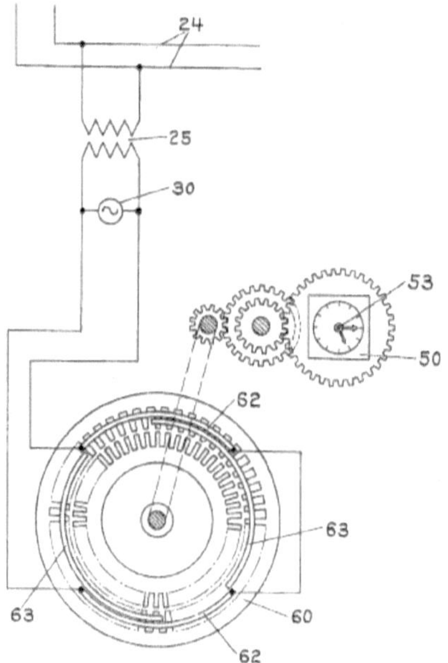

Figure 2-1: Dicke's Vernier motor patent [25]

Similarly, to the doubly-salient machine topology, the Vernier reluctance topology consists of one air-gap separating two ferromagnetic components – a rotor and a stator. Both components have a slotted structure, which creates the alternating air-gap permeance effect, and no permanent magnets. Excitation is provided by current in the windings situated on the stator. The difference in the number of slots on the two components is ultimately what brings about the permeance maxima and minima. This topology was recognised as an attractive option for direct-drive applications due to its high-torque at low-speed functionality. Like modern Vernier machines, the air-gap magneto-motive force (MMF) is modulated by the variable permeance due to the slotted components. Reluctance torque is thus generated [26] [27]. It was recognised here already that a different number of teeth on the rotor (T_r) and the stator (T_s) produced a rotating magnetic field of p_m poles such that:

$$T_s - T_r = \pm n p_m \tag{2.1.1}$$

A rotating minimum reluctance path forms in which the flux follows the path of least reluctance. The rotor naturally seeks to minimise the reluctance between the teeth tips (through the air-gap) to facilitate flux flow. See how rotor and stator teeth can never be perfectly aligned in Figure 2-2. The result is the rotation of the rotor.

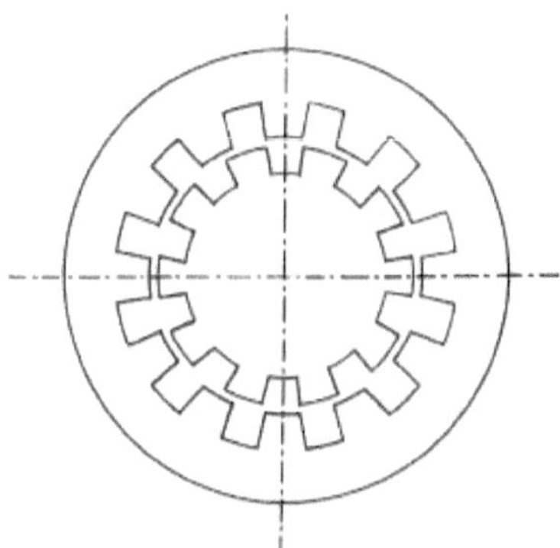

Figure 2-2: Vernier Reluctance motor [26]

Shown in Figure 2-2 is a VRM with 12 stator slots and 10 rotor slots. In the rotor position shown, the stator teeth are facing the rotor slots in the vertical axis, and the rotor and stator teeth are aligned in the horizontal axis. Therefore, in this rotor position, the permeance is maximised along the horizontal axis and minimised along the vertical axis.

If the rotor moves one-half of its own slot pitch, the teeth will align in the vertical axis, and the rotor teeth will align with the stator slots in the horizontal axis. The axis of maximum permeance is now in the vertical axis and the axis of minimum permeance is now in horizontal axis. Therefore, a rotor movement of one-half of a slot pitch displaces the permeance axis by 90^0.

Now, suppose a magnetic field is rotating in the machine air-gap. Whenever the field rotates 90^0, the rotor will rotate one-half of a slot pitch. And one revolution of the magnetic field will correspond to two slot-pitches of rotor movement. If a rotating magnetic field is setup in the air-gap, the rotor will rotate slowly, and at a definite fraction of the speed of the rotating magnetic field. The stepping down of the rotor field is the mechanism which steps up the torque enabling this topology to be labelled high-torque low-speed.

The benefits of the Vernier machine were clearly noted early in its developmental stages. Furthermore, even though it is considered an unconventional topology, its structural simplicity and similarity to that of conventional single-gap machines was always prevalent. This was noted in [28]. The paper sets out to clear up the uncertainty surrounding Vernier machine design. To accomplish this the paper explicitly and concisely describes the Vernier machine governing theory and compares it to the unified theory of machines. The paper shows that, like conventional machines, performance parameters such as inductance can be derived for Vernier machines. And similarly, these performance parameters can be used to derive expressions for torque which provide insight into sizing and output specifications for the Vernier machine. Conveniently, the torque is expressed in terms of stator geometry and coefficients determined by the winding configuration.

The paper which pioneered the Vernier permanent magnet machine is [29]. Whilst previous Vernier designs incorporated the reluctance torque mechanism only, this paper integrates torque due to PM excitation into the reluctance topology. The slotted rotor and stator structures of the Vernier reluctance topology are maintained, but the rotor slots are fitted with permanent magnets for excitation as seen in Figure 2-3.

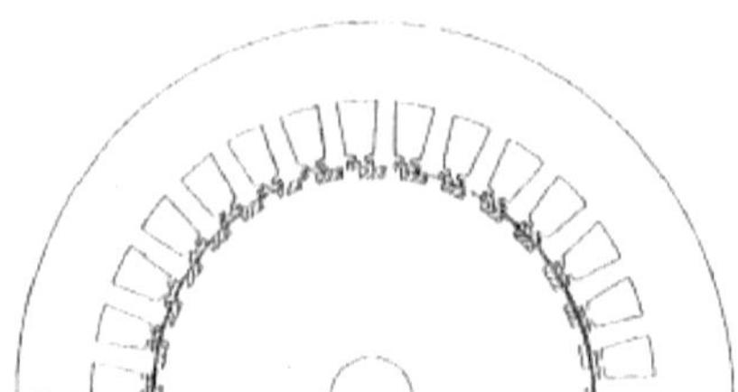

Figure 2-3: Ishizaki VPM design [29]

Although not explicitly noted, the modulation effect is alluded to by the description of the formation of the p_s winding pole-pairs through the interaction of the permeance pulsations and the permanent magnets as illustrated in Figure 2-4.

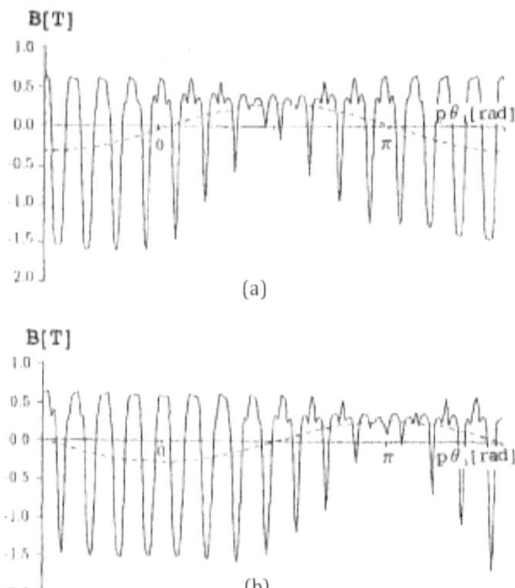

Figure 2-4: Flux density distributions [29] at (a) the starting point and (b) one quarter of a rotor slot pitch later

In Figure 2-4, the solid line represents the no-load air-gap flux density, and the broken lines represent the p_s winding pole-pair flux density component. In Figure 2-4b the rotor has moved one quarter of a rotor slot pitch. Evidently, the winding flux density component has moved by $\frac{\pi}{2}$ electrical radians. Explicitly speaking, the winding component has moved by one pole-pair for the movement of one rotor slot pitch. This is the first empirical evidence of rotor and stator fields rotating at different angular speeds.

2.2 Vernier Design and Theory Survey

The first notable contribution to Vernier PM machine design was made in 2000 by Thomas Lipo and Akio Toba [30], which yielded 102 citations and 1 patent. The authors clearly outlined the fundamentals of Vernier operation. Whilst preceding papers focused on the analysis of the Vernier PM machine, this paper provided a method which breaks up the machine finite element model into an elementary domain thereby providing a design tool for Vernier machines. This elementary domain, shown in Figure 2-5, is a basic building block consisting of one tooth, one slot, and one pole-pair. The structural dimensions of these components are the design parameters used to maximise torque and thus ultimately design the machine. Ultimately, this method cuts simulation time whilst providing guidelines for which parameters need to be maximised during the design process.

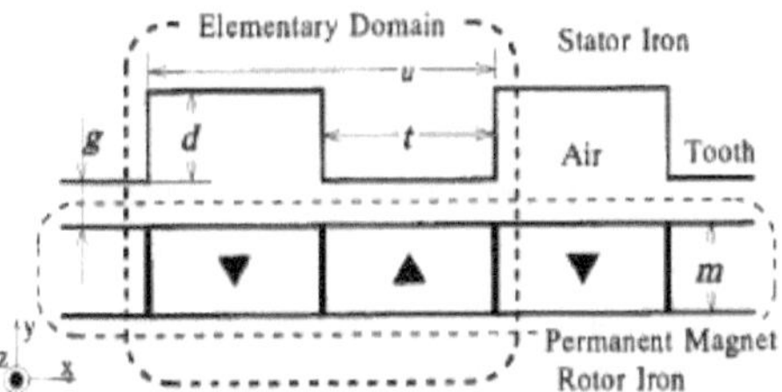

Figure 2-5: Lipo and Toba's Vernier elementary domain [30]

In 2014 Thomas Lipo, this time along with Byungtaek Kim, did it again. The paper "Operation and Design Principles of a PM Vernier Motor" provides invaluable insight into Vernier theory and the so-called "Vernier contribution" [14]. The authors highlight the lack of a clear understanding of the power capability of Vernier machines. More specifically, they note that the air-gap permeance was poorly defined and the result was that the analytical understanding of Vernier machines was poor in comparison to other machines. This paper attempts to bridge this knowledge gap. The Vernier air-gap flux density is derived:

$$B(\theta,\theta_m) = B_0 \cos[p_r(\theta-\theta_m)] - \frac{B_1}{2}\cos\left[(Z \pm p_r)\theta \mp p_r\theta_m\right] \qquad (2.2.1)$$

It is shown to consist of three flux density waves of different pole-pairs: p_r, $Z + p_r$, and $Z - p_r$. Either of the three waves can synchronise with a corresponding winding MMF wave having the same number of pole-pairs. The back-EMF equation for Vernier machines is derived and it is found that it may be written in terms of machine parameters and furthermore, the back-EMF of a conventional PM machine. The two back-EMFs are thus explicitly defined and compared. In the case of the conventional PM machine, a single term is noticed:

$$e_{ph}(t) = k_{1q}N_{ph}D_gL_{stk}\omega_m B_0 cos\left(p_s(\theta_m-\theta_{ph})\right) \qquad (2.2.2)$$

Whereas, two terms are evident in the back-EMF VPM expression:

$$e_{ph}(t) = k_{1q}N_{ph}D_g L_{stk}\omega_m\left(B_0 \mp \frac{p_r}{2p_s}B_1\right)cos\left(p_s(p_r\theta_m \mp p_s\theta_{ph})\right) \tag{2.2.3}$$

Equivalently, equation (2.2.3) can be written in terms of machine geometry:

$$E_{ph} = \frac{2\sqrt{2}B_r}{\pi\mu_r}k_{1q}N_{ph}D_g L_{stk}\omega_m\frac{g_m}{l_g}(K_{conv} \mp K_{add}) \tag{2.2.4}$$

Where K_{conv} is the coefficient for conventional PM machines:

$$K_{conv} = 1 - 1.6k_\beta\frac{\pi D_g}{6p_s q l_g}c_0^2 \tag{2.2.5}$$

And K_{add} is the additional coefficient which takes into account the Vernier contribution:

$$K_{add} = k_\beta c_0\frac{\pi D_g}{p_s l_g}\left(1 - \frac{1}{6q}\right)\left(\frac{0.78}{0.78 - 2c_0^2}\right)\sin(1.6\pi c_0) \tag{2.2.6}$$

The coefficient k_β is the slope of a nonlinear function β which will be expanded upon later. The other parameters are geometry parameters, namely; air-gap diameter D_g, air-gap length l_g, and ratio of slot opening to slot pitch c_0. The number of slots/pole/phase q and winding pole-pairs p_s also feature.
The derived equations and relationships are verified by simulating two VPMs and one PM machine of equal sizes and comparing the results.

The paper rejects the popular notion that the poor power factor in Vernier machines is due to large amounts of leakage flux. This is supported by the fact that the EMF is not greatly reduced in accordance with the large leakage flux. Furthermore, the close correlation between derived formulae and the FEA supports the rejection of the flux leakage hypothesis. The author postulates that because of the machines high operating frequency, the impedance (specifically, reactance) inherently overtakes the EMF. An analytical derivation of the power factor reveals that indeed the power factor is inversely proportional to the gear ratio (and thus Vernier frequency) of the machine. The gear ratio (or pole ratio), the ratio of rotor pole-pairs to winding pole-pairs, is given by:

$$G_r = \frac{p_r}{p_s} \tag{2.2.7}$$

Lipo and Kim go a step further in their design of a Vernier machine for a variable speed application [31] in which conventional machine performance parameters are compared to their Vernier counterparts. This comparison is assessed through the practical suitability for a washing machine application. The machine is designed to function in two modes of operation – wash and spin. The wash mode required 10Nm at 530RPM while the spin mode required 6Nm at 1200RPM. To achieve this an existing conventional PM machine design was used as a foundation for the design of their Vernier prototype. They show that the sizing and design parameters of a Vernier motor can be derived from the performance parameters of a conventional PM machine; where the Vernier parameters are linearly proportional to their conventional counterparts.

The Vernier back-EMF (E_{ver}) is shown to be directly proportional to that of the conventional PM machine (E_{con}) via a proportionality constant K_E:

$$K_E = \frac{E_{ver}}{E_{con}} = \frac{2}{\sqrt{3}} \frac{g'_{con} l_{m,ver}}{g'_{ver} l_{m,con}} \frac{K_{add}}{K_{con}} \tag{2.2.8}$$

If the same air-gap length and magnet thickness is used in both Vernier and conventional machines then:

$$K_E = \frac{2}{\sqrt{3}} \frac{K_{add}}{K_{con}} \tag{2.2.9}$$

Similarly, the Vernier air-gap reactance ($X_{g,ver}$) and conventional reactance ($X_{g,con}$) can be expressed as:

$$K_X = \frac{X_{g,ver}}{X_{g,con}} = \frac{(6q-1)}{0.9c_{Zp,con}} \frac{g'_{con}}{g'_{ver}} \frac{p_{w,con}}{p_{w,ver}} \tag{2.2.10}$$

Where $c_{Zp,con}$ is the ratio of rotor pole-pairs to winding pole-pairs in a conventional machine; which is 1 by definition of a conventional PM machine. If the same air-gap length and number of winding pole-pairs is used in both Vernier and conventional machines then:

$$K_X = 6q - 1 \tag{2.2.11}$$

Considering the Vernier principle in equation (1.2.1) and that $q = \frac{Z}{6p_s}$ for a 3-phase machine, it can be shown that the gear ratio is also equivalent to: $G_r = \frac{p_r}{p_s} = 6q - 1$, which yields an interesting result:

$$K_X = \frac{p_r}{p_s} \tag{2.2.12}$$

This indicates that the air-gap inductance in a Vernier machine is proportional to the Vernier ratio. And because the number of rotor pole-pairs in a Vernier machine is always greater than the number of

winding pole-pairs, it is safe to conclude that the air-gap inductance is inherently larger in Vernier machines compared to conventional PM machines.
Similarly, the Vernier terminal voltage (V_{ver}) can be written in terms of the conventional voltage (V_{con}) and machine performance parameters:

$$k_V = \frac{V_{ver}}{V_{con}} = k_V \sqrt{\frac{1 + \left(\frac{K_X}{K_E}\right)^2 \left(\frac{I_{max} X_{s,con}}{E_{b,con}}\right)^2}{1 + \left(\frac{I_{max} X_{s,con}}{E_{b,con}}\right)^2}} \quad (2.2.13)$$

The analytical study indicates that the air-gap inductance and back-EMF of a Vernier machine has a strong dependence on the number of slots/pole/phase (q) as well as the number of winding pole-pairs (p_s).

In this paper, the Vernier machine had 1.75 times the back-EMF, 3 times the synchronous reactance and required 2.5 times the terminal voltage than that of a conventional PM machine with the same stack length, air-gap diameter and number of turns/phase. The reduction in machine size to meet torque requirement is evidence the Vernier topology can provide higher power density than conventional machines. The downside of this benefit is that the end-turns increased, and coupled with a shorter stack, reduced back-EMF and required a larger terminal current. Therefore, although the deduction of a Vernier design from a conventional PM design is indeed interesting, it may be better to design a Vernier machine from scratch.

A more in-depth study of the slot and pole combination on performance is conducted by Leilei Wu, et al in [32]. FEA is used to investigate the effect that split ratio, slot-opening width and PM thickness along with gear ratio and winding pole-pairs has on torque capacity, ripple, cogging and power factor. Essentially, this paper brings a trade-off to light. The study indicates that torque increases with pole-ratio as seen in Figure 2-6.

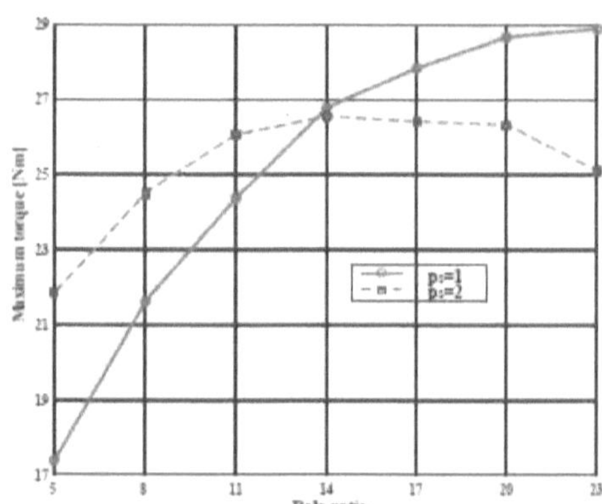

Figure 2-6: Torque vs Pole ratio study conducted in [32]

But noting that the pole-ratio is defined as the ratio of PM pole-pairs to winding pole-pairs, the ratio can only be increased by adding more permanent magnets if the winding pole-pairs is chosen to be one.

Accordingly, additional permanent magnets will yield more leakage flux and depreciate the power factor as seen in Figure 2-7 below. Recall that Lipo, et al. refuted the notion that high leakage flux yields poor power factor in Vernier machines. This is one of many discrepancies in Vernier machine research.

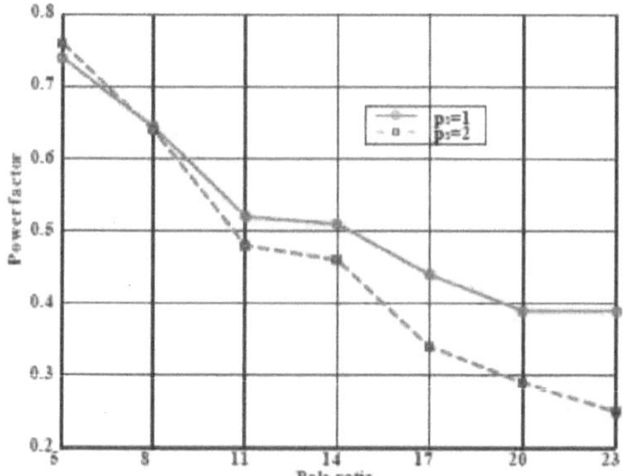

Figure 2-7: Power factor vs Pole ratio study conducted in [32]

The pole-ratio is looked at more closely in [16] where its effect on back-EMF total harmonic distortion (THD) is investigated. This is shown graphically in Figure 2-8. The effect of pole-ratio and tooth width ratio on back-EMF in Vernier machines is analysed. Comparisons between back-EMF waveforms of conventional and Vernier machines are also drawn. This paper mathematically confirms that sinusoidal back-EMF is an inherent quality of Vernier machines. Furthermore, it shows that, unlike in conventional machines, integral-slot full-pitch concentrated winding Vernier machines can also produce sinusoidal back-EMF.

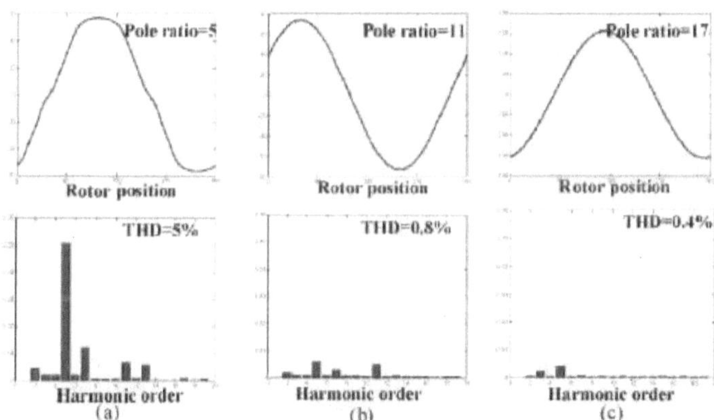

Figure 2-8: A study of the effect of Pole ratio on harmonic quality conducted in [16]

Another inherent characteristic of the Vernier topology is low torque ripple. The mathematics behind this is analysed in [15]. An instantaneous torque expression is analytically derived and thereafter split into nominal and ripple components:

$$T_e = T_{avg} + T_{ripple} \tag{2.2.14}$$

$$= 1.5\pi p_r r_g L_{stk} I_{max} \left\{ F_{c1} \left[N_{\frac{p_r}{p_s}} P_0 + \left(N_{\frac{Z-p_r}{p_s}} + N_{\frac{Z+p_r}{p_s}} \right) \frac{P_1}{2} \right] \right.$$

$$\left. + \sum_{\substack{i=6k\pm1 \\ k=1,2,3\ldots}} iF_{ci} \left[N_{i\frac{p_r}{p_s}} P_0 + \left(N_{\frac{ip_r-Z}{p_s}} + N_{\frac{Z+ip_r}{p_s}} \right) \frac{P_1}{2} \right] \right\}$$

Note that the ripple components exist at the slot harmonics, which are effectively integer multiples of the gear ratio for $i \neq 1$. Additionally, consider that the harmonic parameter magnitudes (F, N and P) are inversely proportional to the harmonic order. Therefore, it is easily deduced that higher order harmonics yield lower magnitude torque ripples. Conversely, the air-gap flux density harmonic orders of $Z + p_r$, $|Z - p_r|$ and p_r are the working harmonics which synchronise with the armature MMF and produce stable torque. Whereas, the other flux density harmonics of $Z + ip_r$, $|Z - ip_r|$ and ip_r produce torque ripple. These harmonic flux density pole-pairs have orders of $\frac{Z+ip_r}{p_s}$, $\frac{|Z-ip_r|}{p_s}$ and $\frac{ip_r}{p_s}$. The number of PM pole-pairs and slots is much higher than the number of winding pole-pairs. Conventional machines have ripple harmonics at ip_s. This is summarised in Table 2-1 below. It follows that the Vernier harmonic orders are much larger than the PM MMF harmonic order i and thus the harmonic amplitudes are much smaller in VPM machines than in conventional PM machines.

Table 2-1: Harmonic order comparison in Vernier and Conventional PM machines [15]

	Harmonic Order	**Working Harmonic Pole-pairs**	**Ripple Harmonic Pole-pairs**
Conventional PM machine	$i = 6k \pm 1$	p_s	ip_s
Vernier PM machine	$\frac{Z+ip_r}{p_s}$, $\frac{\lvert Z-ip_r\rvert}{p_s}$ and $\frac{ip_r}{p_s}$	$Z + p_r$, $\lvert Z - p_r\rvert$ and p_r	$Z + ip_r$, $\lvert Z - ip_r\rvert$ and ip_r

The implication is that the torque ripple is significantly lower in VPM machines. The authors confirm this graphically by comparing the harmonic spectra of a 12-slot 1-winding-pole-pair (and thus an 11-rotor-pole-pair) Vernier machine with a 12-slot 1-winding-pole-pair conventional machine as seen in Figure 2-9 below.

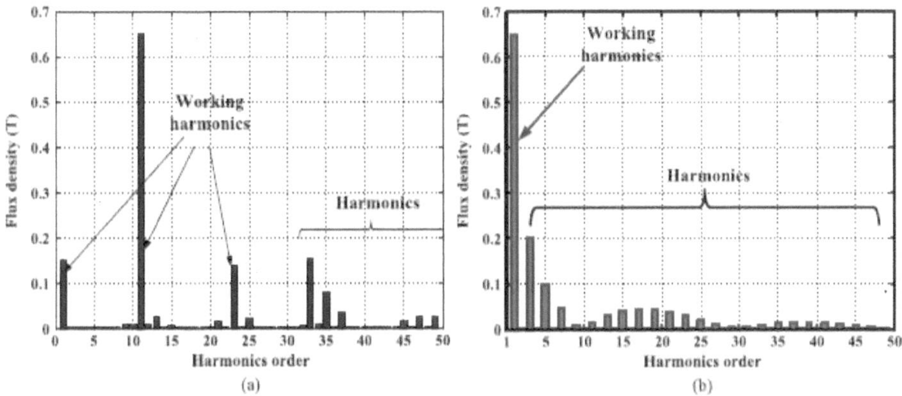

Figure 2-9: Flux density spectra in (a) Vernier and (b) Conventional PM machines [15]

Evidently, although the air-gap in a Vernier machine possesses a much richer flux density harmonic spectrum, the torque ripple in Vernier machines is much lower compared to conventional PM machines. The flux density orders which produce torque ripple are much larger than those in conventional machines.

Nevertheless, the outstanding downside of the Vernier topology is its poor power factor. This was recognised in the aforementioned Vernier reluctance designs and explicitly analysed in [21]. Solutions to this problem are few. Even though this issue is acknowledged throughout Vernier literature, the focus of literature is usually on the Vernier benefits such as appreciable power density. The consensus is that although poor power factor requires a larger rated converter, the Vernier topology still deserves credit for its remarkable suitability for low-speed high-torque applications. The governing theory of power factor will be discussed in Section 3.6. What follows is a treatise on the various Vernier topologies which will provide insight into how some designers seek to mitigate the power factor disadvantage.

2.3 Vernier Machine Topologies

This chapter reviews the various Vernier machine topologies, describing the various structures and operating principles.

2.3.1 Surface mounted – Single air-gap

At first glance, the surface mounted Vernier permanent magnet (SMVPM) machine closely resembles a conventional surface mounted PM machine. The difference is that the number of PM poles does not equal to the number of winding poles by the Vernier principle. Nonetheless, the structural components are the same; a ferromagnetic stator wound with copper windings, radially magnetised permanent magnets mounted on the surface of a ferromagnetic rotor yoke and one air-gap as seen in Figure 2-10.

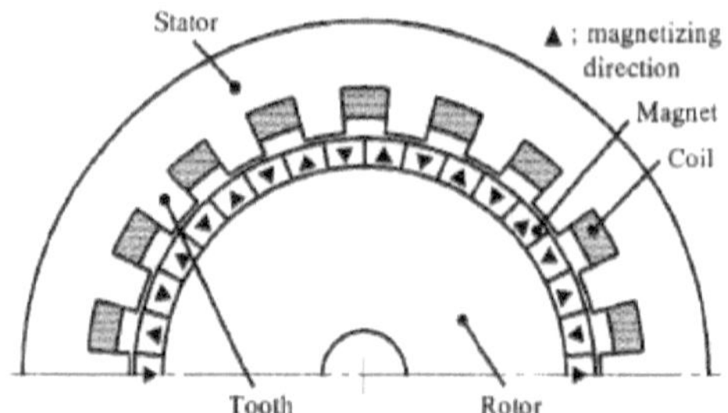

Figure 2-10: SMVPM machine configuration [11]

Operating on the magnetic gear principle, the variable air-gap permeance modulates the MMF setup by the PM rotor. The variable air-gap permeance is set up by the stator teeth itself. The result is an air-gap flux density with two additional slot harmonics which can be synchronised with the armature winding MMF. The consequence of this synchronisation is an augmented power density which is larger in comparison to conventional PM machines. However, although attractive in its simplicity, the surface mounted VPM suffers from high leakage flux and poor power factor.

2.3.2 Dual-Stator Spoke Type

The poor power factor problem of the SMVPM machine is remedied in the Dual-stator configuration. The double-stator Vernier machine topology alleviates the low power factor issue by reducing the harmonic leakage flux and thus also increasing power density [22] [23] [24]. The Dual-stator configuration is depicted in Figure 2-11.

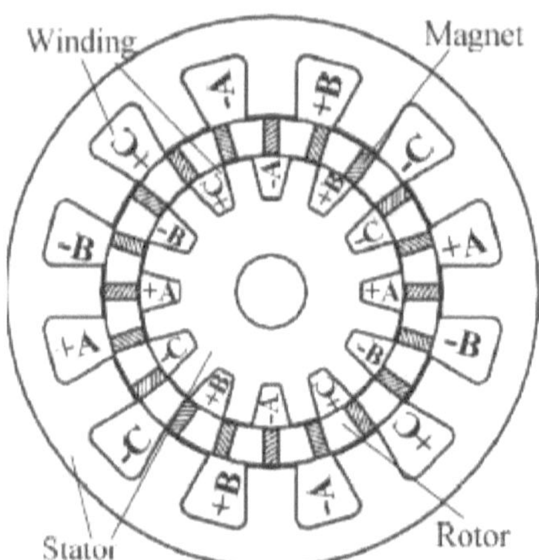

Figure 2-11: Dual-stator Spoke-type [22]

The double-stator topology increases power density by 50% more than that of single-rotor Vernier machines. This is accomplished by employing two air-gaps and two stators, one on either side of the rotor. This setup completes the magnetic circuit thereby reducing flux leakage, increasing power density and improving power factor as shown in Figure 2-12.

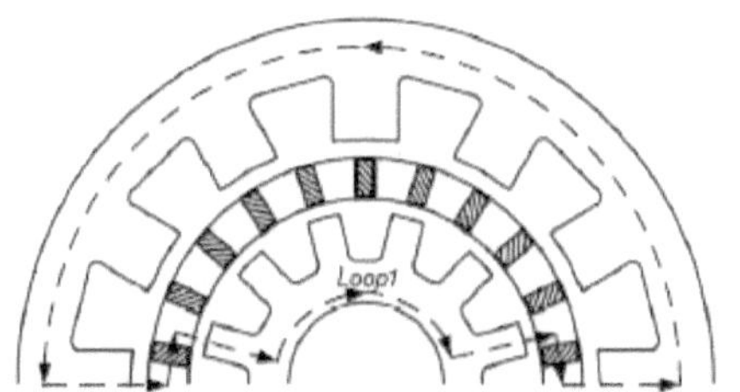

Figure 2-12: Flux loop encompassing the two stators [24]

The necessity of the dual-stator configuration was noted by Lipo and Kim, et al [24], [23]. Besides the poor power factor of the single air-gap configuration, it was discovered the single-gap spoke-type configuration eliminated the Vernier effect altogether by introducing an oscillating magnetic potential in the pole-pieces between the permanent magnets [23]. This is resolved by utilising the dual stator configuration and spoke-type configuration in combination.

In fact, this combination of a spoke-type rotor in a dual-stator topology appears to be in favour among Vernier machine designers [22], [24], [33], [23], [34], [35]. The consensus is that torque density is augmented in the following ways:

- Spoke-type provides the flux focusing effect which yields higher flux density;
- Spoke-type produces a salient effect thus output torque includes magnet and reluctance torque components;
- Dual-stator structure minimises leakage flux.

Conversely, the anisotropy of the spoke-type rotor in Vernier machines is noted and refuted in [33]. Indeed, the d-axis and q-axis inductance would be unequal. However, the Vernier topology uses a relatively large amount of rotor PM poles. Thus, the slot opening, and/or the tooth width is larger than the rotor pole-pitch. This scenario is unique to the Vernier topology. Therefore, the stator teeth can be considered an "anisotropic filter", where the stator teeth attenuate the anisotropy of the rotor. The Vernier spoke topology thus has a negligible saliency ratio.

2.3.3 Vernier Machines with permanent magnets on both Stator and Rotor

An innovative topology is the pole-splitting topology which employs permanent magnets on both the rotor and the stator [36]. Consider the 2014 design in Figure 2-13. The stator core consists of salient teeth which split into the flux modulating pieces (FMPs). The rotor is slotted with permanent magnets situated in the slots. The alternating PM and FMP structure of the rotor is termed a "consequent pole" configuration. The rotor FMPs modulate the field due to the stator permanent magnets and the stator FMPs modulate the field due to the rotor permanent magnets. Consequentially, the two modulated fields couple to produce a comparatively high-performance machine. This is technique is called the "mutual modulation effect". However, the topology falls short in that it fails to utilise the "dead spaces" between ferromagnetic teeth on the stator. The dead spaces are necessary to provide the slotting effect required for modulation. Unfortunately, they also provide a sink for flux loss.

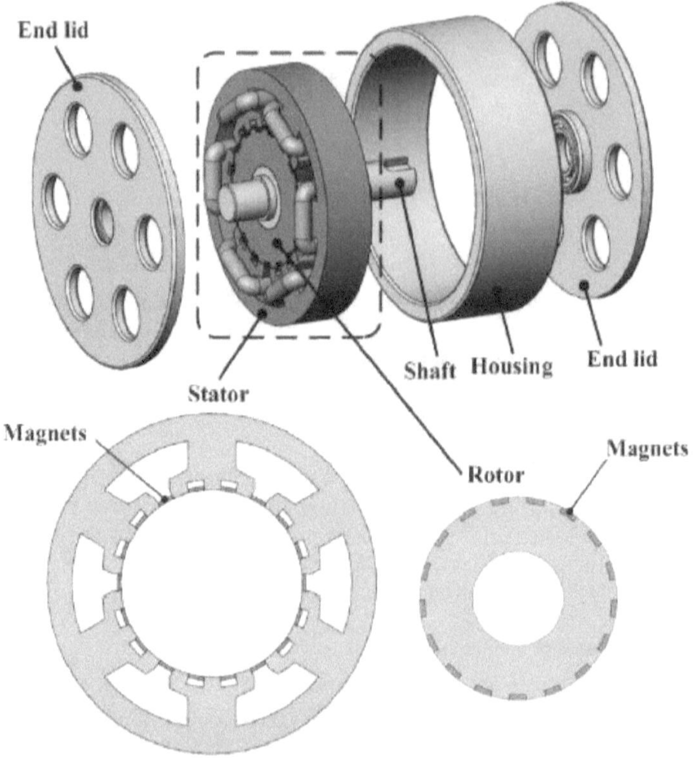

Figure 2-13: Pole-Splitting PM Vernier machine configuration [36]

The topology in Figure 2-14 employs Halbach arrays to mitigate PM flux leakage and enhance flux density [37]. Flux leakage is reduced by using the poly magnetic effect of the Halbach array PM configuration. The Halbach arrays are situated on the stator so as to supplement the winding MMF. Halbach arrays are implemented here by embedding a radially magnetised PM sandwiched between two tangentially magnetised permanent magnets. This produces a flux-focusing effect. The result being that a larger and more stable torque is generated. Moreover, this topology improves upon the conventional pole-splitting topology by incorporating the mutual modulation effect as well.

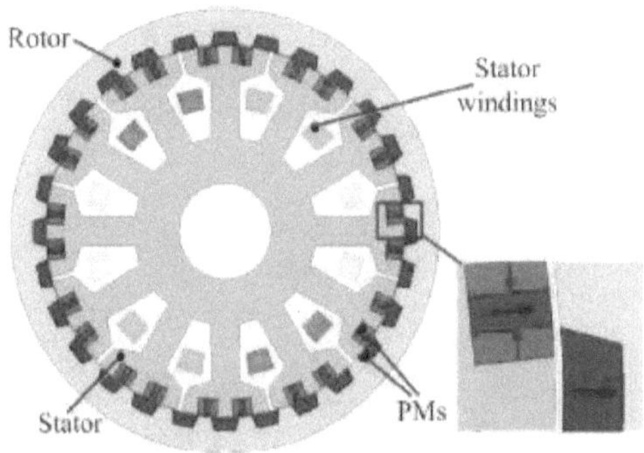

Figure 2-14: Halbach array PM configuration [37]

The negative effect of the dead spaces is alleviated in the Halbach configuration due to its flux focusing effect. Nevertheless, a Halbach array uses excessive amounts of PM material in comparison to the other topologies considered in this book. This is a significant downside.

2.3.4 Superconducting Vernier Machines

Superconducting machines are attractive in that they do not possess permanent magnet material, rather the field poles are set up by the superconductors themselves. Accordingly, superconducting machines will be smaller in size in comparison to conventional PM machines. There exist two types of superconducting machines: High Temperature Superconducting (HTS) in Figure 2-15, and Low Temperature Superconducting (LTS) machines in Figure 2-16.

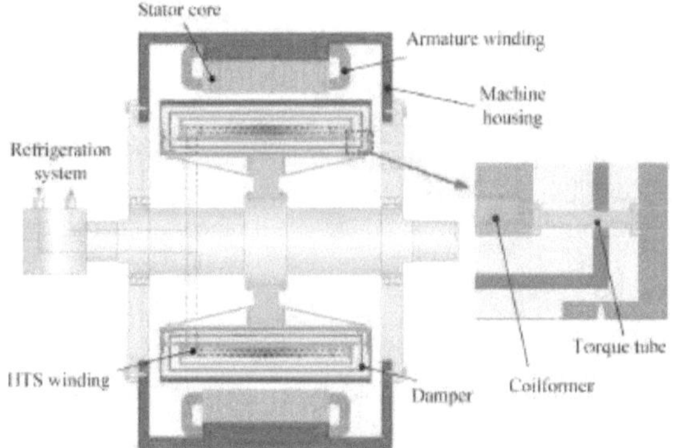

Figure 2-15: HTS Vernier machine assembly [38]

The superconductors are placed on the rotors and thus produce the field poles. These conductors are cooled to below zero temperatures, in the low Kelvin range [39]. At these temperatures conductor resistance is essentially zero. The result is a conductor capable of conducting large currents with copper losses much lower in comparison to that of a conventional PM machine.

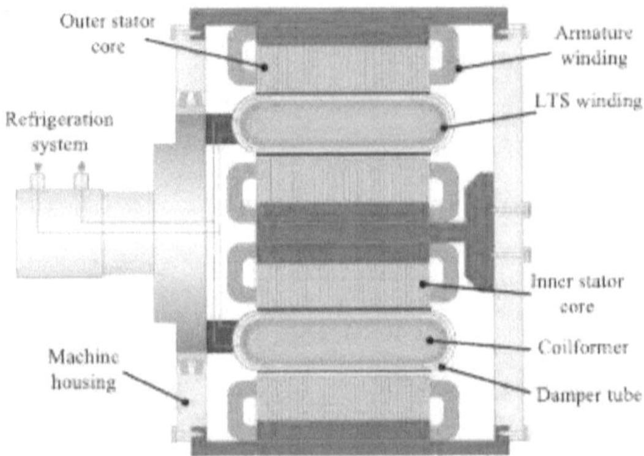

Figure 2-16: LTS Vernier machine assembly [40]

At a temperature below the superconducting material's so-called critical temperature, a pure superconductor will expel an external magnetic field [41]. The critical temperature for a LTS Vernier machine is in the region of 4.2K [40], whereas it can be as high as 138K in an HTS machine [38]. At these temperatures the Meissner effect comes into play. The Meisnner effect constitutes the expulsion of an external magnetic field by a material in its superconductive state. Consequently, the resultant magnetic field is comparatively strong.

Superconducting Vernier machines are designed to incorporate the benefits of superconducting machines and Vernier machines. The characteristic low inductance (and thus high short-circuit current) of superconducting machines is balanced out by the notoriously high inductance of Vernier machines.

2.3.5 Linear Vernier Machines

Linear motors are used in any application requiring linear actuation. Textile machines and systems using magnetic levitation (such as trains) are applications where linear motors are desired [42]. The Vernier linear motor can offer high thrust force at low-speeds which is ideal for linear actuating applications. The surface PM linear Vernier motor consists of permanent magnets arranged on the surface of the mover (armature) whilst the windings are situated on the stator [43] as shown in Figure 2-17.

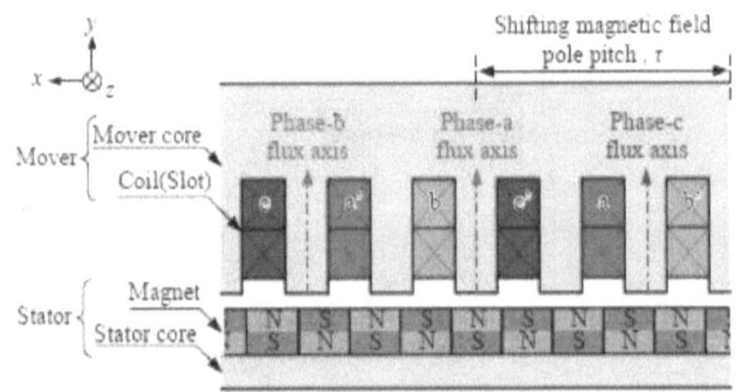

Figure 2-17: Surface mounted linear Vernier PM machine configuration [43]

Similarly, to the classical radial surface mounted PM Vernier machine, the linear version adheres to the Vernier principle, equation (1.2.1) and is wound with a different number of winding poles to rotor PM poles. In its most basic form, the surface mounted linear Vernier motor consists of permanent magnets on the stator and windings mounted on the armature. A popular variation of the linear Vernier topology is the Linear Vernier hybrid (LVH) motor [44]. In this variation, the permanent magnets are situated on the stator teeth surfaces and the coils wound in the stator slots. In this format, it is the armature windings which set up a rotating MMF in the air-gap. The stator is composed of modular U-shaped cores with a coil wound around each primary tooth as seen in Figure 2-18. The armature (or mover) is a ferromagnetic core with salient teeth, and performs the modulation of the air-gap flux density. Thus, it is the armature MMF that is modulated by the air-gap permeance and not the PM MMF (as it is in radial Vernier machines), which is stationary. As a result, the flux density harmonics produced are $(Z \pm p_s)$ and not $(Z \pm p_r)$.

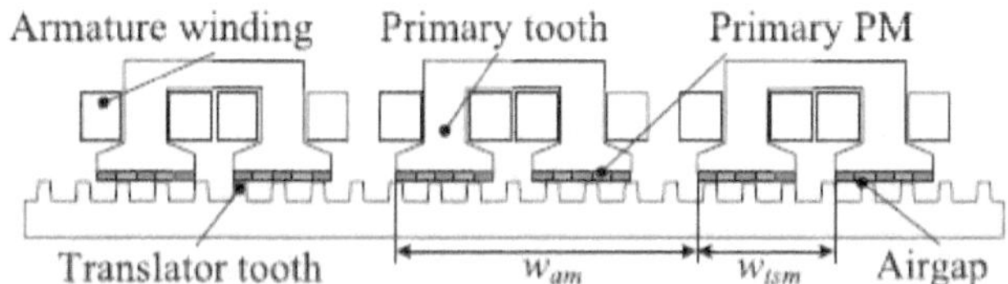

Figure 2-18: Linear Vernier hybrid machine configuration [44]

Another variation is the Linear Primary Permanent Magnet Vernier machine (LPPMV) shown in Figure 2-19. The major difference between the LVH and LPPMV topologies is the stator core. In the LPPMV topology the stator consists of a single laminated core piece instead of the many U-shape modules of the LVH. The role of the translator in this case is to modulate the PM MMF in respect to the magnetic gearing principle.

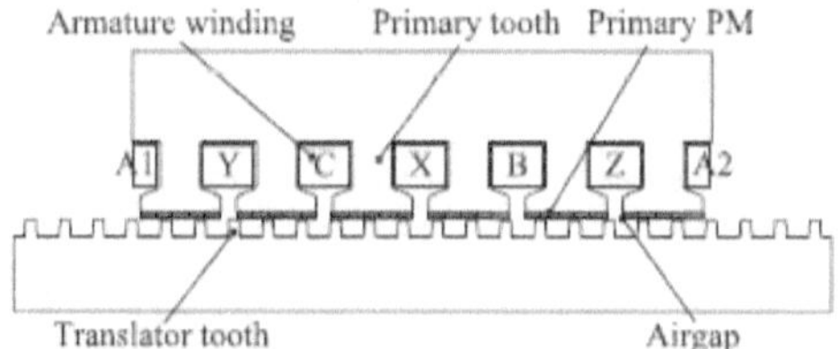

Figure 2-19: Linear primary permanent magnet Vernier machine configuration [44]

2.3.6 Axial flux Vernier Machines

The compact sandwich structure of axial flux VPMs makes it ideal for in-wheel EV applications. The dual rotor axial flux topology employs two rotors with the stator sandwiched in between shown in Figure 2-20 [45].

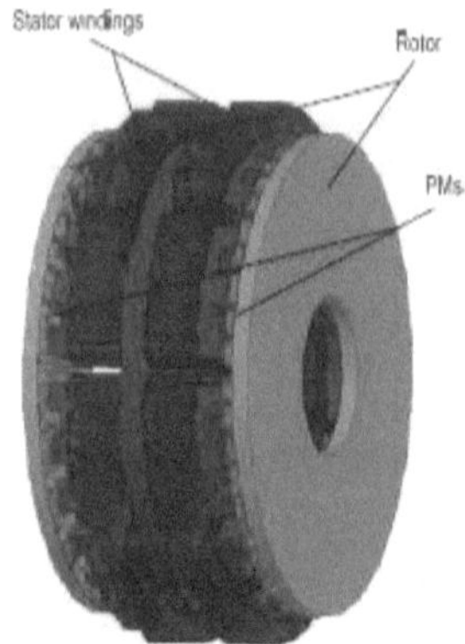

Figure 2-20: Dual rotor axial flux PM Vernier machine configuration [45]

In electric vehicle applications it is preferable to use electric motors which possess a high flux weakening capability to facilitate speed control. The machine flux weakening capability is determined by a large synchronous inductance. A large synchronous inductance is an inherent property of Vernier machines and thus the axial flux Vernier topology is especially suitable for EV in-wheel applications [46] [45].

Zou, Qu, et al. proposed a dual-rotor toroidal winding axial-flux VPM [47] as shown in Figure 2-21. The usage of toroidal windings greatly decreases the end-turn length.

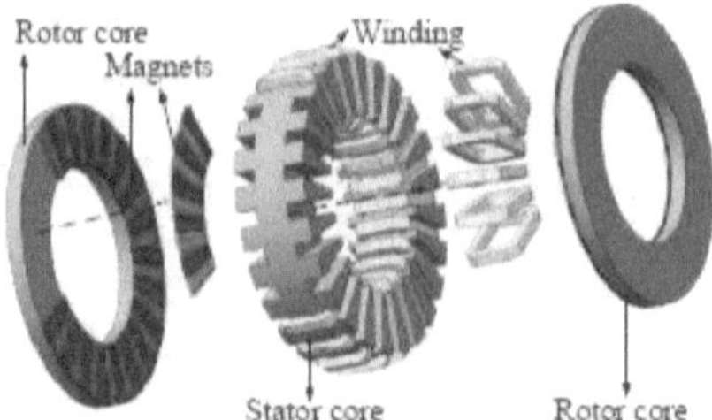

Figure 2-21: Dual rotor toroidal winding axial flux Vernier PM machine configuration [47]

Similarly to other Vernier topologies, the same design parameters are studied; slot opening, pole ratio and PM thickness. Furthermore, it is shown that the Vernier governing equations for axial flux topologies are consistent with other Vernier topologies. This was confirmed using FEA. Each parameter was varied using FEA in terms of torque and compared against the analytical expectations. Results showed good correlation between FEA and theory.

Zhao, Lipo, et al. proposed a dual-stator axial-flux spoke-type PM Vernier motor (DS-AFSPMVM) [48]. The dual-stator axial-flux Vernier topology consists of a single rotor sandwiched between two stators as shown in Figure 2-22. It was hypothesised that this topology would provide all the benefits of the axial flux configuration in terms of high torque density, compact construction and shape flexibility. Additionally, the topology would provide all the benefits of the spoke-type configuration in terms of PM utilisation and high flux density. The rotor is sandwiched between two stators which were unaligned by two-and-a-half slot pitches. In doing so, the flux is guided through the entire machine instead of only half. Furthermore, the unaligned stators have the same effect as skewing, consequently the cogging torque was reduced. 3D FEA simulations showed that this design achieved a power factor of 0.89. This is a significant accomplishment for a Vernier machine with a high pole-ratio of 17.

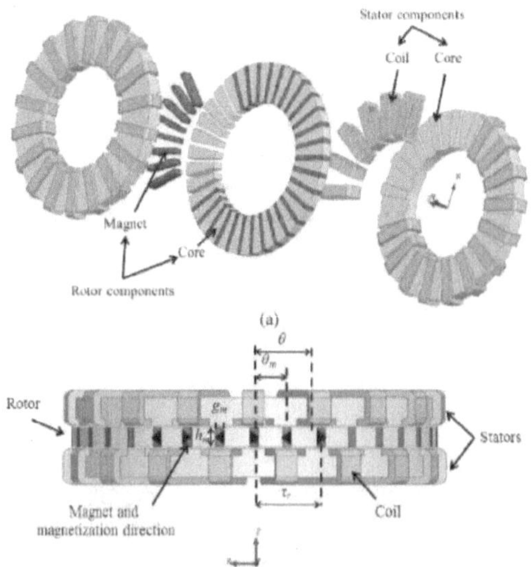

Figure 2-22: DSAFSPMVM configuration [48]

2.3.7 Dual Excitation Vernier Machines

In [49], the authors propose a novel dual-excitation VPM (DE-VPM). The authors motivate the DE-VPM by explaining that the secondary stator makes use of the vacant space inside the primary stator. The rotor has two sets of permanent magnets – on the yoke inner- and outer-surfaces. The dual-excitation VPM structure is shown in Figure 2-23.

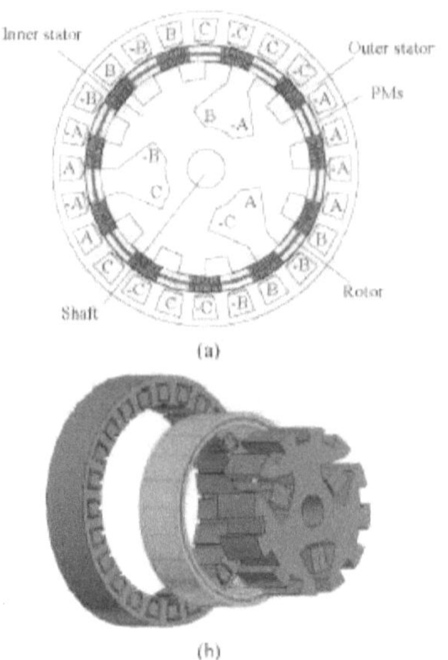

Figure 2-23: Dual-excitation Vernier PM machine structure [50]

This unique design is a combination of a fractional slot concentrated winding (FSCW) conventional PM machine and a Vernier machine. It consists of a rotor nestled between an outer stator and inner stator. The interaction of the rotor with the outer stator is analogous to that of a FSCW conventional PM machine, whereas rotor interaction with the inner stator yields behaviour similar to that of a VPM machine. The inner and outer stator windings are independent of each other thereby providing control flexibility. Both sets of windings can be used simultaneously or independently. The authors believe that the benefits of both types of machines are incorporated into one compact design.

Interestingly, two different flux density waveforms were observed in either air-gap. The inner air-gap flux density was modulated due to the Vernier effect. The outer air-gap flux density resembled that of a conventional PM machine. The back-EMF in either winding had similar magnitudes as well.

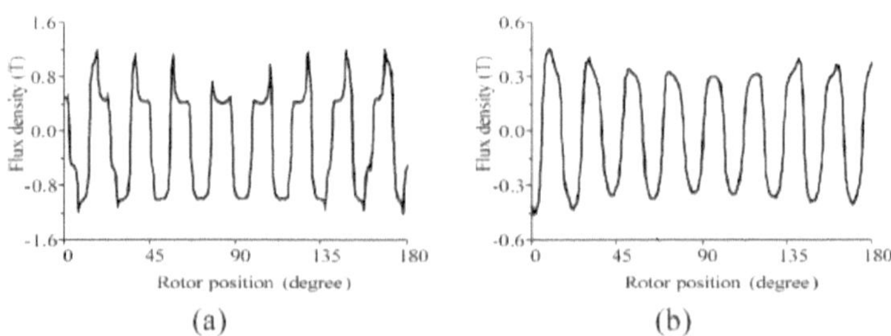

Figure 2-24: DE-VPM flux density waveforms - (a) inner air-gap and (b) outer air-gap [50]

2.3.8 Vernier Machines for Electric Vehicle Applications

The low-speed high-torque functionality of the Vernier topology makes it ideal for electric vehicle (EV) and wind energy conversion system (WECS) applications [51]. Additionally, its relative structural simplicity allows for outer-rotor and axial-flux configurations to be implemented expediently. The compact sandwich structure of axial-flux configurations makes it ideal for in-wheel EV applications; and the outer-rotor configuration enables the hub and spoke facilitation directly to the rotor.

In [52], a variable reluctance Vernier motor (VRVM) is proposed for a hybrid electric vehicle application. The Vernier aspect is selected due to its greater torque density compared to conventional synchronous reluctance machines. The variable reluctance aspect is chosen over a switched reluctance mechanism so that the design can be driven by a three-phase inverter. In addition, this choice saves cost as it does not require expensive permanent magnets.

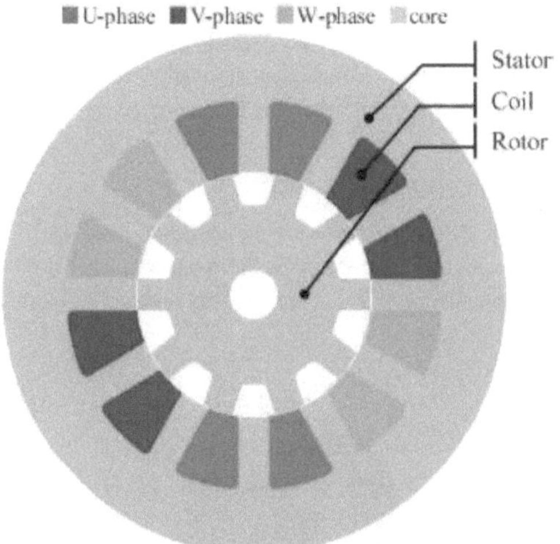

Figure 2-25: Variable reluctance Vernier motor for a hybrid EV configuration [52]

Furthermore, additional endeavours have been made to make a Vernier machine even more suitable for EV applications. In [53], the authors propose an outer-rotor fault-tolerant VPM for an EV application. The fault-tolerance of the machine is ensured by employing fractional-slot concentrated windings as this winding configuration does not have overlapping phases. Ratio of self-inductance to mutual-inductance is 0.42% which is lower than older fault-tolerant designs.

In [12], the authors propose a flux-regulatable VPM for an EV application. Explicitly termed a "dual-magnet Vernier memory machine" (DM-VMM), this topology incorporates the flux-memorisable concept and the Vernier principle. Outer rotors are employed to facilitate direct coupling of rotor to wheel rim in an in-wheel application. Essentially, the paper constitutes a comparison between two structurally slightly different models within this topology.

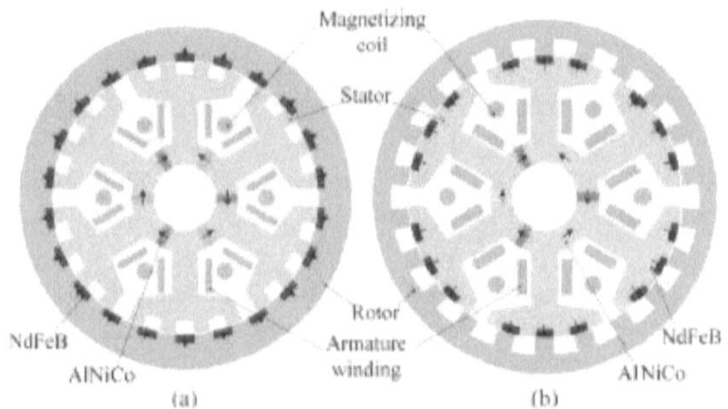

Figure 2-26: Flux regulatable VPM configurations – (a) rotor surface mounted permanent magnets and (b) stator teeth mounted permanent magnets [12]

The flux controllability is achieved by magnetising and demagnetising AlNiCo magnets using a temporary current pulse. The AlNiCo magnets are embedded in the stator yoke in a spoke-type array. Their low coercivity makes them suitable for easy magnetisation and demagnetisation. These auxiliary AlNiCo magnets facilitate large torque production in the low-speed region.

2.3.9 Vernier Machines for Wind Applications

In [13] an outer-rotor VPM for a direct-drive wind application is proposed. The authors claim that the magnetic gearing effect of Vernier machines makes them particularly useful for compensating for the low wind speed and high operating generator speed. This inherently eradicates the speed matching problem that wind turbines have.

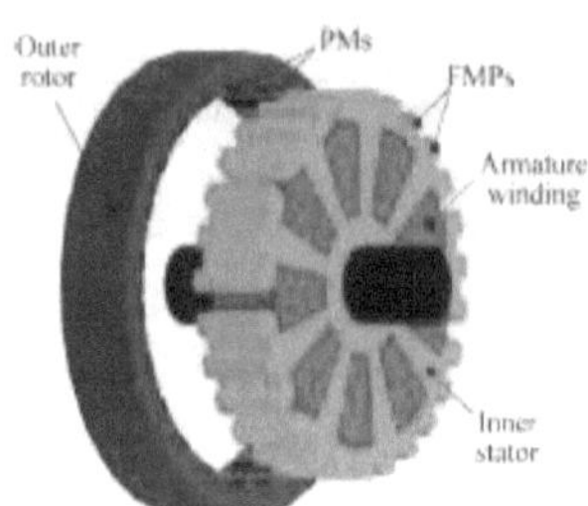

Figure 2-27: Outer-rotor Vernier PM machine configuration [13]

The authors recognise that a Vernier machine has the same capabilities of an integrated magnetic gear machine, without the mechanical complexities.

In [54], the authors describe the necessity of a speed multiplier in a wind turbine drive train. By highlighting the flaws of a mechanical gearbox, they were able to emphasise the benefits of a magnetic speed multiplier (the Vernier machine and magnetic gear family). Subsequently, a theory and design comparison of a Vernier machine, magnetic gear and integrated magnetic gear machine is conducted. The authors conclude that although the initial capital cost of magnetic speed multipliers is larger than that of the mechanical gearbox, the lack of maintenance on magnetic systems will outweigh the high maintenance cost of mechanical systems in the long-term.

3. Theory Development

The following chapter mathematically describes the governing principles behind the Vernier machine and some classical PM machine parameters. The difference between PM Vernier machine theory and conventional PM machine theory will become apparent in this chapter. In conventional machine theory, the performance parameters are characterised solely by the number of pole-pairs p. However, in PM Vernier machines, the characterisation of parameters is split between the number of PM pole-pairs, p_r, and the number of winding pole-pairs, p_s. The flux density is characterised by the number of PM pole-pairs, while winding parameters such as inductance and winding factor are characterised by the number of winding pole-pairs.

3.1 Air-gap Flux Density

Consider the Fourier series of the rotor PM MMF, where i is the order of the working harmonic which exists at odd harmonics only:

$$F_{PM}(\theta_s, t) = \sum_{i=1,3,5,..} F_i \cos(i p_r \theta_s - i\omega t) \tag{3.1.1}$$

Where $F_i = \frac{4}{\pi}\frac{B_r}{\mu_r \mu_0} h_m \sin\left(i\alpha\frac{\pi}{2}\right)$ is the ith Fourier coefficient, α is the ratio of PM arc to pitch, ω is the electrical velocity of the rotor, h_m is the PM thickness in the direction of magnetization, B_r is the PM remanent magnetism and p_r is the number of PM pole-pairs on the rotor. The slotting of the stator teeth creates a variable permeance in the air-gap. This slotting effect is the mechanism by which the air-gap flux density is modulated, thus introducing new order harmonics in the air-gap flux density distribution. In conventional machines the slotting effect is accounted for by use of Carter's coefficient. However, for Vernier machines, Carter's coefficient is unsuitable. A more detailed analysis of the slotting effect for Vernier machines is required to fully describe its operating principle. Henceforth, the conformal mapping method is employed to describe the permeance function coefficients [55]:

$$P(\theta_s) = P_0 + (-1)^n P_1 \cos(Z\,\theta_s) \tag{3.1.2}$$

In which, Z is the number of flux modulating pieces (or slots in this case) and θ_s is the position on the stator. Only the constant P_0 and fundamental permeance harmonics P_1 are taken into account as the higher order harmonics are much smaller in comparison when the slot opening ratio, $\frac{b_0}{\tau}$ is above 0.5 [15]. The conformal mapping method is used to derive the following coefficients of which the detailed derivation is given in Appendix A. The constant permeance coefficient is given by:

$$P_0 = \frac{\mu_0}{g'}\left(1 - 1.6\beta\frac{b_0}{\tau}\right) \tag{3.1.3}$$

The fundamental permeance coefficient is given by:

$$P_1 = \frac{\mu_0}{g'}\frac{4}{\pi}\beta\left[0.5 + \frac{\left(\frac{b_0}{\tau}\right)^2}{0.78125 - 2\left(\frac{b_0}{\tau}\right)^2}\right]\sin\left(1.6\pi\frac{b_0}{\tau}\right) \tag{3.1.4}$$

The parameter, β , is a non-linear function of slot geometry and is given by:

$$\beta = 0.5 - \frac{1}{2\sqrt{1 + \left(\frac{b_0}{2g'}\right)^2}} \qquad (3.1.5)$$

As can be seen from the above expressions, the effectiveness of the modulating effect is governed by the parameters; $\frac{b_0}{\tau}$ the ratio of slot opening to slot pitch, and $\frac{b_0}{g'}$ the ratio of slot opening to magnetic air-gap length.

The air-gap flux density distribution can thus be derived by the product of the permeance function and the PM MMF expression:

$$B_g(\theta_s, t) = F_{PM}(\theta_s, t)P(\theta_s)$$
$$= \sum_{i=1,3,5,..} P_0 F_i \cos(ip_r\theta_s - i\omega t) + \sum_{i=1,3,5,..} P_1 F_i \cos(ip_r\theta_s - i\omega t)\cos(Z\theta_s) \qquad (3.1.6)$$

Then, using $\cos(\mathrm{u})\cos(v) = \frac{1}{2}[\cos(u - v) + \cos(u + v)]$, the resulting flux density is:

$$B_g(\theta_s, t) = \sum_{i=1,3,5,..} P_0 F_i \cos(ip_r\theta_s - i\omega t)$$
$$+ \sum_{i=1,3,5,..} \frac{P_1 F_i}{2} \cos((Z - ip_r)\theta_s + i\omega t) \qquad (3.1.7)$$
$$+ \sum_{i=1,3,5,..} \frac{P_1 F_i}{2} \cos((Z + ip_r)\theta_s - i\omega t)$$

The resultant expression for the air-gap flux density indicates the presence of three flux density waves each having different numbers of pole-pairs (and thus spatial orders); ip_r, $(Z - ip_r)$ and $(Z + ip_r)$. Each wave rotates at the same electrical frequency ω. It is worth noting that the first term in equation (3.1.7) can be regarded as the component found in classical machine theory used for conventional PM machines. Whilst, the other two terms are produced by the modulation effect of the stator slots.

3.2 Torque Production

Synchronous rotation in a PM electrical machine is caused by the interaction of two magnetic fields – the air-gap flux density due to PM excitation and the armature winding MMF. Stable torque is only generated by synchronising terms in the field expressions having the same spatial orders. The conditions for stable torque production are thus [56]:

- Stator and rotor field have the same number of poles. Fields which have different number of poles do not interact with each other;
- Stator and rotor field are rotating at the same speed.

Torque ripple results when the stator field and rotor field have the same number of poles but rotate at different speeds. The armature MMF of the three-phase windings is given by the product of the winding function and the input current:

$$F_a = N_a(\theta_s)i_a(t) + N_b(\theta_s)i_b(t) + N_c(\theta_s)i_c(t)$$
$$= \frac{3}{2}\sum_{\substack{j=6k+1\\k=0,1,2,..}} N_j I_{max}\cos(jp_s\theta_s - \omega t) + \frac{3}{2}\sum_{\substack{j=6k-1\\k=0,1,2,..}} N_j I_{max}\cos(jp_s\theta_s + \omega t) \quad (3.2.1)$$

Essentially, based on the governing principle of Vernier machines, equation (1.2.1), it is only necessary to subsequently consider terms which share spatial periods in the air-gap flux density and armature MMF. Terms in these two aforementioned expressions which do not share spatial orders do not synchronise and thus do not contribute to net torque.

Thus considering the air-gap flux density expression given by equation (3.1.7), the stator MMF harmonics, jp_s, will synchronise with either the ip_r or the $Z \pm ip_r$ air-gap flux density slot harmonics such that:

$$jp_s = \begin{cases} ip_r \\ Z \pm ip_r \end{cases} \quad (3.2.2)$$

This is the essence of the Vernier principle. Synchronisation is ensured by choosing the number of winding pole-pairs such that equation (1.2.1) is true. Therefore, considering only the fundamental and slot harmonics the stator MMF in equation (3.2.1) can alternatively be expressed as:

$$\begin{aligned} F_a(\theta_s, t) = \frac{3k_{w1}F_{c1}}{2}\Bigg\{&\cos[p_s\theta_s - (\omega t - \alpha)] \\ &+ \frac{(-1)^n}{Z/p_s - 1}\cos[(Z - p_s)\theta_s + (\omega t - \alpha)] \\ &+ \frac{(-1)^{n-1}}{Z/p_s + 1}\cos[(Z + p_s)\theta_s - (\omega t - \alpha)]\Bigg\} \end{aligned} \quad (3.2.3)$$

If the jp_s armature fields synchronises to the $Z - ip_r$ rotor fields, the interaction of the two fields produces a stable torque which can be derived from the field energy in the air-gap [11]:

$$\frac{co - energy}{volume} = \frac{1}{2}BF \quad (3.2.4)$$

Where F is the sum of the PM and armature MMF's and B is the sum of the air-gap and armature flux densities. The volume in question is the air-gap volume. The derivation follows from:

$$T = \frac{p_s\beta_w L_{st}}{\pi}\int_0^{2\pi}\left\{\frac{\partial}{\partial\theta_m}\left(\frac{1}{2}BF\right)\right\}d\theta_s \quad (3.2.5)$$

Where β_w is the winding pole-pitch, L_{st} is the stack length and θ_m is the rotor position.

$$T = \frac{p_s \beta_w L_{st}}{\pi} \int_0^{2\pi} \left\{ \frac{\partial}{\partial \theta_m} \left(\frac{1}{2} (F_{PM} P + F_a P)(F_{PM} + F_a) \right) \right\} d\theta_s$$
$$= \frac{p_s \beta L_{st}}{\pi} \int_0^{2\pi} P(F_{PM} + F_a) \frac{\partial}{\partial \theta_m} F_{PM} \, d\theta_s \quad (3.2.6)$$

From equations (3.1.1), (3.1.2), (3.1.7), and (3.2.3) and setting α to $\pi/2$ (in equation (3.2.3)) the maximum torque per ampere can be obtained:

$$T = \frac{3\sqrt{2}}{\pi} \beta L_{st} k_{w1} N_{ph} I_a \left(\frac{p_r}{p_s} B_1 + B_0 \right) \quad (3.2.7)$$

From equation (3.2.7) it is evident that there is an additional component to the torque in a Vernier machine. This component is the manifestation of the Vernier effect. This is due to the synchronisation of the PM MMF slot harmonic to the armature MMF. The term can be scaled by the factor p_r/p_s theoretically infinitely. However, in practice this comes at a cost which will be explored later.

3.3 Vernier Back-EMF Formulation

Consider a three-phase machine with phases A, B, C; where phases B and C are displaced from phase A by $\frac{2\pi}{3}$ and $\frac{4\pi}{3}$ respectively. From winding function theory, the winding function of stator phase A winding is given as:

$$N_a(\theta_s) = \sum_{j=1,3,5,..} N_j \cos(j p_s \theta_s) \quad (3.3.1)$$

$N_j = \frac{2}{j\pi} \frac{N_{ph}}{p_s} k_{wj}$ Is the jth harmonic of one phase armature winding function, N_{ph} is the number of turns per phase and k_{wj} is the winding factor of the jth harmonic. The flux linkage of phase A is the product of the phase A turns per coil per phase and the flux density linking phase A.

$$\lambda_a(\theta_s, t) = \int B_g(\theta_s, t) \cdot dA$$
$$= R_g L_{st} \int_0^{2\pi} B_g(\theta_s, t) \cdot N_a(\theta_s) \cdot d\theta_s$$
$$= R_g L_{st} \int_0^{2\pi} \left\{ \sum_{i=1,3,5,..} \sum_{j=1,3,5,..} B_0 N_j \cos(i p_r \theta_s - i\omega t) \cos(j p_s \theta_s) + \sum_{i=1,3,5,..} \sum_{j=1,3,5,..} B_1 N_j \cos((Z - i p_r)\theta_s + i\omega t) \cos(j p_s \theta_s) + \sum_{i=1,3,5,..} \sum_{j=1,3,5,..} B_1 N_j \cos((Z + i p_r)\theta_s - i\omega t) \cos(j p_s \theta_s) \right\} \cdot d\theta_s \quad (3.3.2)$$

At this point, define $B_0 = P_0 F_i$ as the slot-less flux density harmonic component which is found in conventional PM machines. While $B_1 = \frac{P_1 F_i}{2}$ is defined as the slotting flux density harmonic component which is unique to Vernier machines.

Then, expanding using: $\cos(u)\cos(v) = \frac{1}{2}[\cos(u - v) + \cos(u + v)]$

$$\begin{aligned}\lambda_a(\theta_s, t) = R_g L_{st} \sum_{i=1,3,5,..} \sum_{j=1,3,5,..} B_0 N_j \frac{1}{2} \int_0^{2\pi} \{\cos[(ip_r - jp_s)\theta_s - i\omega t] \\ + \cos[(ip_r + jp_s)\theta_s - i\omega t]\} \cdot d\theta_s \\ + R_g L_{st} \sum_{i=1,3,5,..} \sum_{j=1,3,5,..} B_1 N_j \frac{1}{2} \int_0^{2\pi} \{\cos[(Z - ip_r - jp_s)\theta_s \\ + i\omega t] + \cos[(Z - ip_r + jp_s)\theta_s + i\omega t]\} \cdot d\theta_s \\ + R_g L_{st} \sum_{i=1,3,5,..} \sum_{j=1,3,5,..} B_1 N_j \frac{1}{2} \int_0^{2\pi} \{\cos[(Z + ip_r - jp_s)\theta_s \\ - i\omega t] + \cos[(Z + ip_r + jp_s)\theta_s - i\omega t]\} \cdot d\theta_s\end{aligned} \tag{3.3.3}$$

Expanding each integral and considering each term in equation (3.3.3) individually:

The first term is due to the linkage of the jp_s winding pole-pairs with the ip_r flux density pole-pairs. This is termed the slot-less component. This is the very same back-EMF contribution found in conventional PM machines.

$$\therefore I_1 = \sum_{i=1,3,5,..} \sum_{j=i\frac{p_r}{p_s}} \pi B_0 N_j \cos(i\omega t) \tag{3.3.4}$$

The second term is due to the linkage of the jp_s winding pole-pairs with the $Z - ip_r$ flux density pole-pairs. This is one of the so-called slotting components. It is a unique feature of the Vernier topology.

$$\therefore I_2 = \sum_{i=1,3,5,..} \sum_{j=\frac{Z-ip_r}{p_s}} \pi B_1 N_j \cos(i\omega t) \tag{3.3.5}$$

The third term is due to the linkage of the jp_s winding pole-pairs with the $Z + ip_r$ flux density pole-pairs. This is one of the other slotting components.

$$\therefore I_3 = \sum_{i=1,3,5,..} \sum_{j=\frac{Z+ip_r}{p_s}} \pi B_1 N_j \cos(i\omega t) \tag{3.3.6}$$

Thus, the flux linkage becomes:

$$\lambda_a(\theta_s,t) = \pi R_g L_{st}\left\{\sum_{i=1,3,5,..}\sum_{j=i\frac{p_r}{p_s}} B_0 N_j \cos(i\omega t) + \sum_{i=1,3,5,..}\sum_{j=\left|\frac{Z\pm ip_r}{p_s}\right|} B_1 N_j \cos(i\omega t)\right\} \tag{3.3.7}$$

The developed back-EMF is defined as the rate of change of flux-linkage with respect to time:

$$e_a(t) = -\frac{d}{dt}\lambda_a(\theta_s,t)$$

$$= -\pi R_g L_{st}\left\{\sum_{i=1,3,5,..}\sum_{j=i\frac{p_r}{p_s}} B_0 N_j \frac{d}{dt}\cos(i\omega t) + \sum_{i=1,3,5,..}\sum_{j=\left|\frac{Z\pm ip_r}{p_s}\right|} B_1 N_j \frac{d}{dt}\cos(i\omega t)\right\}$$

$$= \pi\omega R_g L_{st}\left\{\sum_{i=1,3,5,..}\sum_{j=i\frac{p_r}{p_s}} iB_0 N_j \sin(i\omega t) + \sum_{i=1,3,5,..}\sum_{j=\left|\frac{Z\pm ip_r}{p_s}\right|} iB_1 N_j \sin(i\omega t)\right\} \tag{3.3.8}$$

The winding factors $k_{w\frac{p_r}{p_s}}$ and $k_{w\left|\frac{Z\pm p_r}{p_s}\right|}$ are slot harmonics and are therefore equal in magnitude to the fundamental winding factor. Thus, considering only the working harmonic $i = 1$ and making the necessary substitutions, $j = i\frac{p_r}{p_s}$ and $j = \left|\frac{Z\pm ip_r}{p_s}\right|$ results in a simpler form for back-EMF. This form is used in the sizing analysis.

$$e_a(t) = N_{ph} k_w \omega_m D_g L_{st}\left(B_0 + \frac{p_r}{p_s}B_1\right)\sin(\omega t) \tag{3.3.9}$$

Clearly, the Vernier back-EMF consists of two terms. The first term is the same found in that of conventional machines. The second term is the Vernier contribution. If poles and slots are chosen such that $Z = p_r + p_s$, the terms are summed.

3.4 PM Machine Three-phase Windings

The method of star of slots is used to determine the winding layouts of the machines. The star of slots method consists of an array of spokes which represents the fundamental EMF harmonic induced in one coil-side of a coil. The method which follows here is adapted from [57]. In this paper, the winding configuration will be characterised by the machine number of slots per pole per phase:

$$q = \frac{Z}{2mp_s} \tag{3.4.1}$$

Z is the number of slots, m the number of phases, and p_s the number of winding pole-pairs. Fractional slots are defined by non-integer values for q. While integral slot windings are defined by integer values of q such that $q \geq 1$.

3.4.1 Method of Star of Slots

The star of slots is divided into $2m$ sectors each displaced by $360/2m$ degrees. A sector is characterised as either positive or negative with two sectors belonging to the same phase. The corresponding sectors of a phase are displaced by 180 degrees. The resulting winding layouts are shown in Figure 4-6. The machine periodicity, t , is defined as the greatest common divisor between the number of slots Z and the winding pole-pairs p_s.

$$t = GCD\{Z, p_s\} \tag{3.4.2}$$

The angle between adjacent slots, or the slot pitch, is given in mechanical radians by

$$\alpha_s = \frac{2\pi}{Z} \tag{3.4.3}$$

There are $^{Z}/_{t}$ spokes in the star of slots. Spokes, not to be confused with the slots, represent the EMF phasor which is induced in each coil-side. The angle between the spokes is given by

$$\alpha_{ph} = \frac{2\pi t}{Z} = \alpha_s t \tag{3.4.4}$$

Or in electric radians

$$\alpha_{ph}^e = p_s \alpha_s t \tag{3.4.5}$$

In the star of slots, adjacent slots are displaced by in electrical radians:

$$\alpha_s^e = \frac{p_s}{t} \alpha_{ph} \tag{3.4.6}$$

This study constitutes three-phase machines. Therefore, the star of slots consists of six sectors; positive sectors A+, B+, C+ and negative sectors A-, B-, C-. The number of spokes per phase sector is $Z/2mt$ and the rules of thumb determining the locations of spokes in the star of slots are:

1. If Z/t is even then both the positive and negative sectors of a particular phase have an equal number of spokes
2. If Z/t is odd then the number of spokes in the positive and negative sectors of a particular phase differ by 1.

3.4.2 Winding Factor

In a distributed winding the coils are distributed over several slots. The voltage induced in each coil is thus displaced by the angle between spokes on the star of slot diagram, α_{ph}. The Distribution factor is given by:

$$k_d = \frac{\sin\left(q_{ph}\frac{\alpha_{ph}}{4}\right)}{q_{ph}\sin\left(\frac{\alpha_{ph}}{4}\right)} \tag{3.4.7}$$

Where the number of spokes per phase is

$$q_{ph} = \frac{Z}{mt} \tag{3.4.8}$$

In a short-pitched coil, the actual coil span is less than the pole pitch. The pitch factor can be considered as the ratio of the voltage produced by a short-pitched winding to the voltage produced by a full pitched winding. The pitch factor, which is less than one for short-pitched coils, is:

$$k_p = \sin\left(\frac{\pi p_s\, y_q}{Z}\right) \tag{3.4.9}$$

The actual coil span or coil throw, which is the number of slots which a single winding pole encompasses, is given by

$$y_q = round\left\{\frac{Z}{2p_s}\right\} \tag{3.4.10}$$

Finally, the winding factor is:

$$k_{wj} = k_{pj}k_{dj} \tag{3.4.11}$$

3.5 Inductance

Inductance is a consequence of the armature reaction effect in electrical machines. It is a property of the winding distribution and stimulated by the winding armature MMF rather than the PM MMF [58].

3.5.1 Air-gap Inductance

The air-gap inductance of a single-layer full-pitch sinusoidally distributed winding is given by [31], where g'' is the effective magnetic air-gap length:

$$L_g = \frac{\pi}{4}\frac{\mu_0 N_{ph}^2 L_{st} D_g}{p_s^2 g''} \tag{3.5.1}$$

3.5.2 Slot Leakage Inductance

This inductance is created by the magnetic flux, that traverses the slot area in the circumferential direction, as shown in Figure 3-1. Consider the classical description of inductance [59]:

$$L = \mathcal{P}{N_{tc}}^2 \tag{3.5.2}$$

In the equation (3.5.2), $\mathcal{P}$ is the permeance of the magnetic path and N_{tc} is the number of turns per coil. Hence, the slot leakage inductance can be determined by the slot permeance coefficient. For un-tapered trapezoidal slots, the permeance coefficient of a rectangular slot suffices:

$$P_{slot} = \frac{1}{3}\frac{d_{slot}}{b_0} \tag{3.5.3}$$

This represents the permeance experienced by the flux which runs parallel to the slot in the circumferential direction.

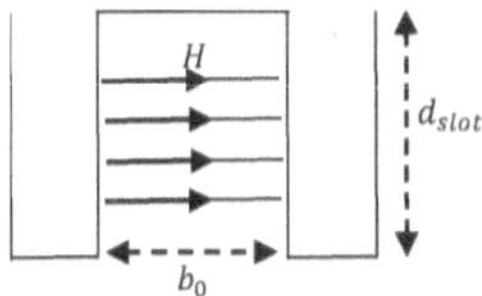

Figure 3-1: Simplified slot geometry for inductance formulae derivations

Following from [60] , the slot inductance becomes:

$$L_{slot} = 2p_s L_{st} P_{slot} {N_{tc}}^2 (4q - N_{sp}) \tag{3.5.4}$$

Where a full pitch coil is given by

$$W = \frac{Z}{2p_s} \tag{3.5.5}$$

And the number of short pitches is

$$N_{sp} = W - y_q \tag{3.5.6}$$

3.5.3 End-turn Leakage Inductance

End-turn leakage inductance is created by the magnetic flux which leaves one slot and before it enters another.

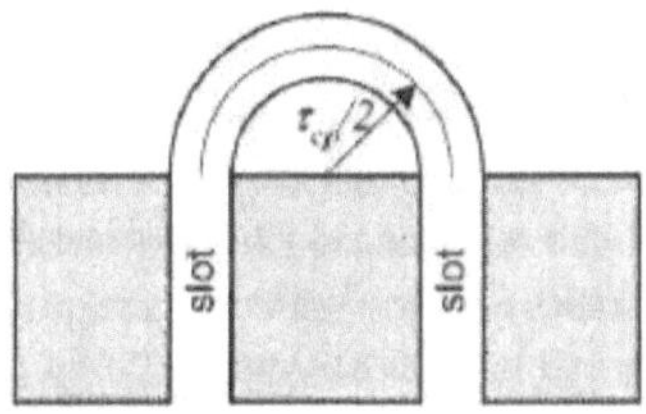

Figure 3-2: End-turn geometry simplification [61]

For $2p_s$ winding poles and two end-turns per pole, the end-turn inductance is:

$$L_{ew} = p_s \mu_0 \tau_{cp} N_{tc}^{\ 2} \ln\left(\frac{\tau_{cp}\sqrt{\pi}}{\sqrt{2A_{slot}}}\right) \tag{3.5.7}$$

Where τ_{cp} is the mean coil pitch of a single coil in milli-meters and A_{slot} is the slot area.

3.5.4 Synchronous Reactance

The total synchronous inductance is the sum of the inductance components:

$$L_{syn} = L_g + L_{slot} + L_{ew} \tag{3.5.8}$$

The synchronous reactance is the sum of the air-gap magnetising inductance and the leakage inductances. In a conventional PM machine, the synchronous reactance is given as:

$$X_{g,con} = \omega_m p_s L_{syn} \tag{3.5.9}$$

For a Vernier machine, where the number winding and rotor pole-pairs differ, the synchronous reactance is:

$$X_{g,ver} = \frac{p_r}{p_s}\frac{\pi}{4}\frac{\mu_0 N_{ph}^2 L_{st} D_g}{p_s g''}\omega_m \tag{3.5.10}$$

The dependence of reactance on pole-ratio is evident:

$$X_{g,ver} = \frac{p_r}{p_s} X_{g,con} \tag{3.5.11}$$

Additional insight can be gained by noting the dependence on the slots/pole/phase q:

$$X_{g,ver} = (6q - 1)\, X_{g,con} \tag{3.5.12}$$

Equations (3.5.11) and (3.5.12) are indeed the conclusions reached by [31] as discussed previously in Section 2.2.

3.6 Terminal Characteristics and Power Factor

In a reactive circuit, the reactive components (inductors and/or capacitors) do not dissipate power. Rather, equal amounts of power are being absorbed and returned to the source [62]. Hence, if the source were a generator and the load purely reactive, it would require (practically) zero mechanical energy to turn the shaft as no power is being used by the load.

3.6.1 Terminal Characteristics in Wind Energy Conversion Systems

For the case of a generator coupled to a converter, reactive power is constantly exchanged between the machine armature circuit and the converter.

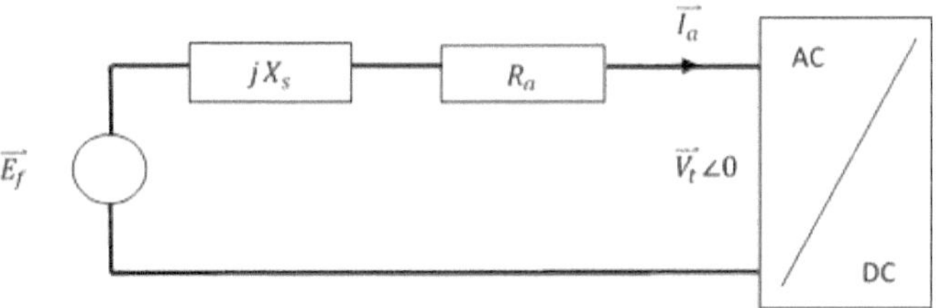

Figure 3-3: Per-phase equivalent circuit of a generator with terminals coupled to a converter

This implies that the converter components must be capable of withstanding nominal real power as well as the exchanged reactive power; and must therefore be sized accordingly. Reactive power and real power are related via the apparent power, which is the complex sum of the two power components:

$$\vec{S} = P + jQ \tag{3.6.1}$$

This can be represented in a vector triangle, such that:

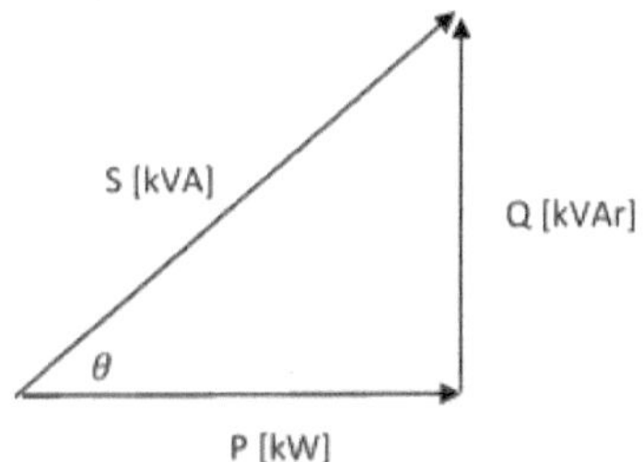

Figure 3-4: Vector representation of equation (3.6.1)

The angle θ is the power angle the cosine of which is called the power factor (PF). Evidently, the amount of reactive power Q that is transmitted back-and-forth between the source and the load is dependent on the machine's power factor. The power factor is in turn dependent on the internal characteristics of the machine; namely the machine winding reactance. The coils which constitute the machine phase windings consume reactive power and is thus termed an inductive load, producing a lagging power factor. The reactive power is the theoretical driving force behind the establishment of the magnetic field in electrical machines.

In the direct-drive wind turbine scheme, the generator would most likely be coupled to an inverter-rectifier combination. Passive diode rectifiers are most commonly used when the project capital cost is relatively low, such as in low power commercial wind turbine systems. A boost chopper is employed to convert the variable output rectifier voltage to constant voltage [63]. However, these rectifiers induce harmonics in the current waveforms which reduces efficiency and contributes to torque ripple [64]. Moreover, it lowers the power factor.

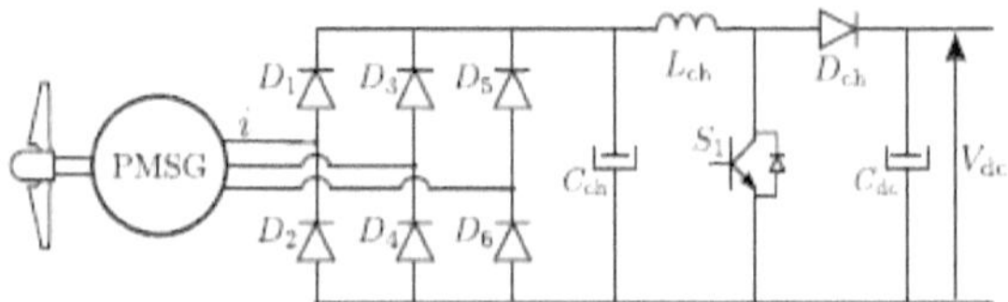

Figure 3-5: **Passive diode rectifier and boost chopper [63]**

Essentially, the DC-link capacitor smooths the rectifier output voltage. The greater the capacitance, the smoother the DC-link voltage. However, this smoothness comes at a cost – greater harmonic density in the current waveform. Harmonic content in the current waveform yields lower efficiency. The current harmonics can be alleviated by increasing the machine terminal inductance. The inductance acts as a passive filter, filtering out parasitic harmonics. However, increasing inductance has a negative implication – it lowers the power factor [64].

Alternatively, a less crude harmonic mitigation strategy would be to implement an LCL filter on the grid-side. The LCL filter enables the use of a voltage source converter (VSC) as an active rectifier. The VSC introduces current harmonics due its high switching frequencies but offers full control of the DC-link voltage, and even the power factor. A more cost-effective harmonic attenuation option is provided by the LCL filter [65].

The phase terminal voltage characteristic of an electrical machine in generator mode is given by:

$$\vec{V_t} = \vec{E_0} - (R_{ph} + jX_{syn})\vec{I_a} \tag{3.6.2}$$

Where $\vec{E_0}$ is the open circuit back-EMF, $\vec{I_a}$ is the winding current and R_{ph} and X_{syn} are the phase resistance and synchronous reactance respectively.

Equation (3.6.2) and Figure 3-3 tell an interesting story. The terminal voltage is determined by the back-EMF and synchronous reactance. As discussed, Vernier inductance is inherently larger than that of conventional machines. Therefore, due to the Vernier machine characteristically high inductance:

1) The synchronous reactance and thus reactive power would be higher;
2) For the same amount of real power, the apparent power would be higher;
3) It follows that the power factor would be low;
4) The terminal voltage would be low.

Herein lies the issue with Vernier machines; it is difficult and impractical to size a converter to a machine that generates low terminal voltage with a high apparent power rating.

3.6.2 Power Factor in Vernier Machines

The proportionality of the pole-ratio to the air-gap reactance (equation (3.5.11)) and the characteristic high frequency of Vernier machines are two factors that exacerbates the power factor situation in Vernier machines. In effect, the synchronous reactance of Vernier machines 'overtakes' the back-EMF contributing towards a low power factor. This is particularly evident in the expression for Vernier power factor:

$$cos\varphi = \frac{1}{\sqrt{1+\left(\frac{p_r}{p_s}\right)^2\left(\frac{X_{g,con}I_{ph}}{E_{ph}}\right)^2}} \tag{3.6.3}$$

This analytically shows that Vernier machines have an inherently poor power factor. Furthermore, there is an apparent trade-off between power density and power factor. The foregoing analysis indicates that there cannot exist a high-power density Vernier machine with a high-power factor. The practical implication being Vernier machines would require converters which have higher apparent power ratings than conventional PM machines.

3.6.3 Power Factor in non-ideal Machines

An interesting approach for machine power factor analysis is adopted in [66]. In electrical machines, saturation of the ferromagnetic material may occur. This saturation leads to non-sinusoidal currents or voltage waveforms. Such harmonic-dense waveforms are unsuitable for determining the power factor from the phase difference between voltage and current waveforms. Traditionally, power factor is defined as the rate of real power P to apparent power S:

$$PF = \frac{P}{S} = \frac{\frac{3}{T}\int_0^T V_i(t)I_i(t)dt}{3V_{RMS}I_{RMS}} \tag{3.6.4}$$

In this case the voltage is purely sinusoidal and is given by $V_i(t) = V_{pk}\sin(\omega t)$. Similarly, the current is $I_i(t) = I_{pk}\sin(\omega t - \phi)$. The angle ϕ is the phase difference between the voltage and current waveforms. The real power is thus:

$$P - \frac{3}{T}\int_0^T V_{pk}sin(\omega t)I_{pk}sin(\omega t - \phi)dt \tag{3.6.5}$$

Then, let $a = \omega t$ and $b = \omega t - \phi$ such that $a - b = \phi$ and $a + b = 2\omega t + \phi$:

$$P = \frac{3}{T}\int_0^T V_{pk}sin(a)I_{pk}sin(b)dt \tag{3.6.6}$$

Finally, use the trigonometric identity $sin(a)sin(b) = \frac{1}{2}\cos(a-b) - \frac{1}{2}\cos(a+b)$:

$$P = \frac{3V_{pk}I_{pk}}{2T}\int_0^T [cos(\phi) - cos(2\omega t + \phi)]dt \tag{3.6.7}$$

It is easily shown that $\frac{1}{T}\int_0^T cos(\phi)dt = cos(\phi)$ and $\frac{1}{T}\int_0^T cos(2\omega t + \phi)dt = 0$. Therefore:

$$P = \frac{3V_{pk}I_{pk}}{2T}\int_0^T cos(\phi)dt$$
$$= \frac{3V_{pk}I_{pk}}{2}cos(\phi) \quad (3.6.8)$$

The RMS values of the waveforms are defined as $V_{RMS} = \frac{V_{pk}}{\sqrt{2}}$ and $I_{RMS} = \frac{I_{pk}}{\sqrt{2}}$. And the power factor is thus:

$$PF = \frac{P}{S}$$
$$= \frac{\frac{V_{pk}}{\sqrt{2}}\frac{I_{pk}}{\sqrt{2}}\cos(\phi)}{V_{RMS}I_{RMS}}$$
$$= \cos(\phi) \quad (3.6.9)$$

This definition is true for purely sinusoidal waveforms where the phase difference is constant whether measured from the waveform zero crossings or the waveform peaks. This is evident in Figure 3-6.

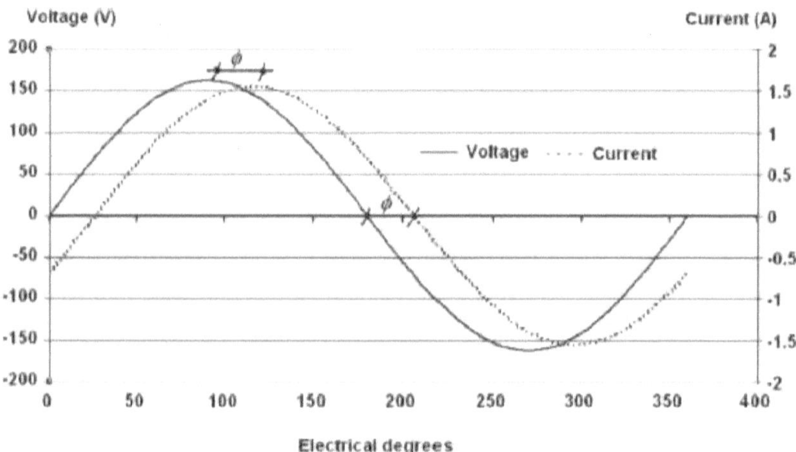

Figure 3-6: The phase difference of purely sinusoidal voltage and current waveforms [66]

However, in reality the current is charged with harmonics. In such a case the phase difference is inconsistent – the phase difference at the zero crossings is inequivalent to the phase difference at the peaks.

The presence of harmonics is unavoidable. Harmonics can arise from converters, reactors and magnetic saturation. Thus, a more realistic definition is required. Consider a purely sinusoidal input voltage and a current consisting of fundamental and third harmonic components; the real power is then:

$$P = \frac{3}{T}\int_0^T V_1 sin(\omega t)[I_1 sin(\omega t - \phi_1) + I_3 sin(3\omega t - \phi_3)]dt \quad (3.6.10)$$

$$= \frac{3}{T}\int_0^T V_1 sin(\omega t) I_1 sin(\omega t - \phi_1)dt + \frac{3}{T}\int_0^T V_1 sin(\omega t) I_3 sin(3\omega t - \phi_3)dt$$

Clearly, the first term in this equation will, similarly to the purely sinusoidal case, yield a power factor of $PF = \cos(\phi_1)$. The second term requires a closer look:
Let $a = \omega t$ and $b = 3\omega t - \phi_3$ such that $a - b = -2\omega t + \phi_3$ and $a + b = 4\omega t + \phi_3$:

$$P_{harm} = \frac{3}{T}\int_0^T V_1 sin(a) I_3 sin(b)dt \tag{3.6.11}$$

Finally, use the trigonometric identity $sin(a)sin(b) = \frac{1}{2}\cos(a-b) - \frac{1}{2}\cos(a+b)$:

$$P_{harm} = \frac{3V_1 I_3}{2T}\int_0^T [cos(-2\omega t + \phi_3) - cos(4\omega t + \phi_3)]dt \tag{3.6.12}$$

It can be shown that: $\frac{1}{T}\int_0^T cos(-2\omega t + \phi_3)dt = 0$ and $\frac{1}{T}\int_0^T cos(4\omega t + \phi_3)dt = 0$. Therefore, it can be deduced that voltage and current waveforms which do not have the same frequency do not contribute towards real power. Hence, only the fundamental harmonic of the current contributes towards active power, such that:

$$P = \frac{V_1}{\sqrt{2}}\frac{I_1}{\sqrt{2}}\cos(\phi_1) \tag{3.6.13}$$

Evidently, to determine the power factor one would have to extract the fundamental current component and its phase difference, relative to the input voltage. Therefore, in the case where the effect of current harmonics are taken into account, the term $\cos(\phi_1)$ is defined as the "displacement power factor".

A more revealing definition of power factor is the "true power factor" [67]. The true power factor takes the effect of the total harmonic distortion (THD) into account. THD is defined as the ratio of the RMS value of the harmonics to the RMS value of the fundamental:

$$THD\% = \frac{\sqrt{\sum_{n=3,5,7,..} I_{nRMS}^2}}{I_{1RMS}} \tag{3.6.14}$$

The true power factor is thus defined as:

$$PF_{true} = \frac{P}{3V_{1RMS}I_{1RMS}}\frac{1}{\sqrt{1 + THD^2}} \tag{3.6.15}$$

This method is suited for FEA in which a voltage source is used to excite the machine windings. In such a scenario, the input voltage is purely sinusoidal whereas the resultant currents contain harmonics.

3.7 Cogging Torque

Cogging torque is due to the interaction of the rotor PM poles and the stator steel teeth. Although there is no simple means to explicitly calculate the cogging torque magnitude, the LCM provides an indication of the degree of cogging torque the machine will exhibit [68]. Strictly speaking, it gives the number of cogging periods per revolution. It is particularly useful in a comparative study. The LCM is the lowest common multiple between the rotor poles and the stator teeth. The machine with the largest LCM will exhibit the lowest cogging torque.

$$f_c = \frac{2p_r Z}{LCM\{2p_r, Z\}} \tag{3.7.1}$$

The analytical expression for cogging torque is derived from the magnetic energy in the air-gap under PM excitation only [69]:

$$W(\theta) = p_r \varphi_{m0} \phi(\theta) \tag{3.7.2}$$

Then the cogging torque is:

$$T_{cog}(\theta) = \frac{\partial W(\theta)}{\partial \theta} = p_r {\varphi_{m0}}^2 \frac{\partial \Lambda_g(\theta)}{\partial \theta} \tag{3.7.3}$$

Where φ_{m0} the no-load MMF due to PM excitation and $\phi(\theta)$ is the flux in one PM pole. Considering that the Vernier topology will use an open slot structure to facilitate the flux density modulation, it is necessary to describe the cogging torque in terms of the consequent variable reluctance. The subsequent theory is adapted from [70]. This is a consequence of the slotting effect of PM machines, which is characterised by a variable reluctance, or permeance.

$$T_{cog}(\theta) = \frac{1}{2} i^2 \frac{dL}{d\theta} - \frac{1}{2} \phi^2 \frac{dR}{d\theta} + Ni \frac{d\phi}{d\theta} \tag{3.7.4}$$

Where the cogging torque term is given by

$$T_{cog}(\theta) = -\frac{1}{2} \phi^2 \frac{dR}{d\theta} \tag{3.7.5}$$

Evidently, the cogging torque is directly proportional to the variable reluctance $\left(\frac{dR}{d\theta}\right)$. In conventional PM machines methods such as stator slot skewing and teeth tapering are used to suppress the variable permeance and thus reduce cogging torque [71] [68]. Skewing minimises the variable reluctance along the axial direction by spreading the slot openings out over the magnet area. Thus, each magnet sees a diminished net reluctance which is almost non-varying as they pass individual slots.

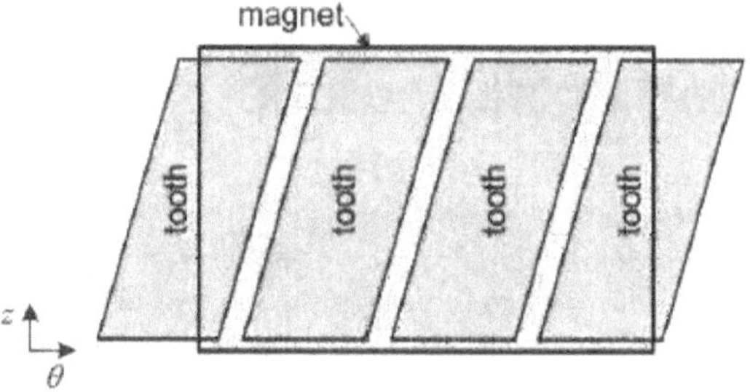

Figure 3-7: Magnet and slotted tooth interface when skewing is implemented [70]

However, as outlined in Section 3.1, the Vernier topology uses the variable permeance to modulate the rotor MMF. Thus, in this sense, it would not be wise to implement skewing or teeth tapering in Vernier machines.

3.8 Copper Losses

The copper losses in the stator windings are usually the largest contributors to energy conversion losses [72]. The copper loss in the windings is proportional to the total length of the windings, hence the length of the windings along the stack and the length of the end-turns needs to be accounted for. To simplify calculations, the end-turns are assumed to be semi-circular. Therefore, the copper loss is subsequently determined. The stator winding per-phase resistance for a DC current is given by [73]:

$$R_{1DC} = \frac{N_{ph} l_{1av}}{a\sigma A_c} \tag{3.8.1}$$

Where N_{ph} is the number of turns per phase, l_{1av} is the average length of a single turn, a is the number of parallel paths, σ is the electrical conductivity for a given temperature ($\sigma \approx 57 \times 10^6\,[S/m]\ at\ 20^oC\ and\ \sigma \approx 47 \times 10^6\,[S/m]\ at\ 75^oC$), and A_c is the conductor cross-sectional area. However, in the case of an AC current, the resistance differs, and the skin-effect can be taken into account. The resistance of the coils along the machine axis and the end-windings are considered:

$$R_{AC} = \frac{2N_{ph}}{a\sigma A_c}(L_{st}k_{1R} + l_{1e}) \tag{3.8.2}$$

Where, the average length of a winding turn is given by $l_{1av} = 2(L_{st} + l_{1e})$ and k_{1R} is the skin-effect coefficient. Where L_{st} is the stack length and l_{1e} is the length of a single end-turn. It can be assumed that the end-turns follow a semi-circular path with the radius equal to half of the mean coil pitch.

For small machines energized with 50-60Hz sources, it is safe to assume that the skin effect is negligible. In fact, R_{AC}/R_{DC} is less than 1.01 in this case [74]. Thus, the actual stator per-phase resistance in this case is given by equation (3.8.1) as this book concerns low-speed applications only. Therefore, the total machine stator copper loss is:

$$P_{cu} = mI_a{}^2 R_{1DC} \tag{3.8.3}$$

3.9 Core losses

Iron core losses in electrical machines consists of eddy current loss and hysteresis loss [75].

3.9.1 Hysteresis Loss

An electrical machine connected to an external AC power source receives energy from the source during the first half-cycle of conduction. Whilst energy is returned to the source during the second half-cycle. The core and winding coils consume more energy than is returned to the source. It is this difference in energy that goes into heating the core. This power loss is called hysteresis loss. The hysteresis loss in electric machine cores is given empirically by:

$$P_h = K_h B_{pk}^{\beta} f \tag{3.9.1}$$

3.9.2 Eddy Current Loss

There is relative motion between the rotating rotor permanent magnets and the stationary stator core. As the flux from magnets cuts the cross section of the stator core, a voltage is induced on the cross-sectional surface due to Faraday's Law. A current thus flows in an enclosed path induced by this voltage. This current is termed eddy currents. The steel core is electrically conductive and has a finite resistance. The resultant i^2R loss heats the core and is empirically quantified by:

$$P_e = K_e B_{pk}^2 f^2 \tag{3.9.2}$$

3.9.3 Excess Loss

In addition to hysteresis and eddy current losses, there exists a smaller component to core loss called excess loss [76]. Excess or anomalous losses are caused by the realigning and rearranging of the core magnetic domains as the steel is magnetised by the PM excitation. Excess loss can be determined as such:

$$P_a = \frac{K_a}{T} \int_0^T \left(\frac{dB}{dt}\right)^{1.5} dt \tag{3.9.3}$$

3.9.4 Total Core Loss

As per convention, hysteresis, excess and eddy current loss are grouped into a single quantity – the core loss:

$$P_c = P_h + P_e + P_a \tag{3.9.4}$$

3.10 Rotational losses

The rotational losses consist of friction losses in the bearings and windage losses caused by the rotating rotor. The friction loss is thus given by [77].

$$P_{fr} = 0.001 k_{fb} m_{rotor} n_s \tag{3.10.1}$$

Where $k_{fb} = 1\ to\ 3$ is the bearing friction coefficient provided by the manufacturer. The windage loss is described as follows for speeds which do not exceed 6000RPM [78]:

$$P_{wind} = 2D_g{}^3 L_{st} n_s{}^3 \times 10^{-6} \tag{3.10.2}$$

The total rotational losses is thus the summation of the two components:

$$P_{rot} = P_{fr} + P_{wind} \tag{3.10.3}$$

3.11 Efficiency

Efficiency is defined as the ratio of output power to input power.

$$\eta = \frac{P_{out}}{P_{in}} \tag{3.11.1}$$

In generator mode the output power is given by electrical power which is generated

$$P_{out} = P_{elec} = 3V_{ph}I_{ph}\cos(\theta) \tag{3.11.2}$$

Whereas the input power is given by the shaft power

$$P_{in} = P_{shaft} = T\omega_m \tag{3.11.3}$$

Equivalently, the efficiency can be described in terms of the machine losses:

$$P_{out} = \frac{P_{out}}{P_{in} + P_{loss}} \tag{3.11.4}$$

Where the total machine losses is given by:

$$P_{loss} = P_{cu} + P_c + P_{rot} \tag{3.11.5}$$

4. Sizing, Design Constraints, and Goals

4.1 Design Methodology

The design methodology employed in this book is as follows

1) Choose the number poles and slots for the four design variations;
2) Determine the winding configuration and layout - one fractional slot configuration and one integral slot configuration;
3) Select the magnet grade and define the initial pole size;
4) Use the magnet specifications to determine the air-gap diameter and stack length;
5) Calculate magnetic loading and determine magnetic circuit geometry;
6) Calculate electric loading and determine slot geometry, taking into account the winding configurations;
7) Design, size and simulate the SM designs;
8) Use same stator and rotor diameter from the SM designs for the spoke designs;
9) Parameterise the magnet sizes in the spoke cases to ensure rated specs are met.

The machine is to have rated power of 6kW at a rated speed of 250RPM based on the load requirements of a typical wind farm and wind speeds in the area. The generator must generate a terminal voltage of 380V line-line; the residential voltage in SA. These are the design targets used in this study. This translates to a rated torque of 230Nm at 10A and generated voltage of 380V line-line. In practice, the generator is connected to a variable-frequency converter, so a frequency constraint need not be employed. No constraints are placed on the sizes of the machines. Rather the machines are sized such that it is ensured that the specified performance targets in Table 4-1 are met.

Table 4-1: Machine Performance Parameters

Rated Power [kW]	6
Rated Speed [RPM]	250
Torque [Nm]	230
Line-line Voltage [V]	380
RMS Current [A]	10

4.2 Poles and Slots

The pole-ratio, equation (2.2.7), is at the core of Vernier machine design theory. Its effect is the proportionality factor which amplifies the slot harmonic component of the air-gap flux density, thereby supplementing the power density. This harmonic component is otherwise negligible in conventional machines and does not increase power density of the machine.

The dependency of power density on the selected pole configuration is therefore quite clear. In light of this theory, one would be inclined to design a Vernier machine with a large number of rotor pole-pairs, thereby maximising the pole-ratio and consequently the power density. However, this will yield excessive leakage flux and diminishing power density. There exists a delicate balance between high pole-ratio to achieve high power density, and the number of permanent magnets that yields high leakage and consequently diminishes power density [32].

In addition to the preceding consideration, the selected rotor poles, winding poles and slots must abide by the Vernier principle (equation (1.2.1)). Furthermore, this study requires two pole and slot combinations such that $q \in \mathbb{Z}$ and $q < 1$ for integral slot and fractional slot configurations respectively. In accordance with the wind generator requirements outlined earlier, both integral and fractional pole combinations must satisfy the following criteria:

- Low cogging torque
- Appreciable power density
- The number of slots must be a multiple of three, as this is a three-phase machine design
- Acceptable power factor

Naturally it will be impossible to meet all of the aforementioned criteria. Trade-offs between certain of these factors exist and thus compromises will be made. Tables B.1 – B.4 in Appendix B were used to select the pole-slot combinations. The frequency is determined from the number of rotor pole-pairs and the rated speed:

$$f = \frac{RPM \times p_r}{60} \tag{4.2.1}$$

Although frequency is not an exact requirement, it is desirable to have a pole and slot combination that yields a frequency of 50 ± 10Hz. This frequency range is based on the fact that silicon steel used in electrical machines is optimised for core losses within this range. This frequency range results in a narrowed down range for the pole and slot options. Furthermore, the fundamental winding factor is restricted to k_w> 0.9, in order to ensure a reasonable induced voltage in the stator windings.

4.2.1 Integral Slot Pole and Slot Selection

For the integral slot case, noting the trade-off between power density and leakage flux, a high pole-ratio is chosen which uses relatively few permanent magnets, hence the 12 slot 22 pole combination was chosen. This combination has q =2. Choosing a combination with q = 1 yields a lower pole-ratio, thereby sacrificing power density. Whereas a value of q greater than 2 utilises too many permanent magnets which would result in higher material cost and excessive leakage flux.

A closer look at the tables in Appendix B highlights how the 12 slot 22 pole combination was chosen. In Table B.1, the q=1 case has too few poles to meet the frequency requirement. The other integral slot cases in Table B.1 do not meet the frequency criterion either. In Section 3.7 it was explained that machines which have a higher LCM between the poles and slots exhibit lower cogging torque. Here this notion is put into practice. In Table B.2, the q=2 case yields an LCM which is lower in comparison to the q=2 case in Table B.1 even though it does meet the frequency criterion. In Table B.3, the q=1, case has a lower LCM than the q=2 case in Table B.1. Moreover, the q=2 case in Table B.3 does not meet the frequency criterion. In Table B.4, the q=1 case does not meet the frequency criterion, nor does it have a greater LCM than the q=1 case in Table B.1. In summary, the 12 slot 22 pole case meets all of the aforementioned criteria and offers appreciable power density due to its high pole ratio.

4.2.2 Fractional Slot Pole and Slot Selection

For the fractional slot case, it was decided to select a slots/pole/phase combination such that q<1. This was decided upon in order to reduce the winding inductance and therefore improve the machine's power factor substantially. Equation (3.5.12) highlights the proportionality of q to reactance, while equation (3.6.3) highlights the inverse proportionality of power factor to reactance.

Table B.1 does not satisfy the q<1 criterion. In Table B.2, two cases meet the q<1 criterion, but underperforms in terms of frequency and winding factor. In Table B.3, q=0.833 meets both the frequency and q<1 criterion. However, the LCM proves inadequate in comparison to the LCM of the q=0.75 case in Table B.4. Therefore, the 18 slot 28 pole combination was therefore chosen. This combination has q = 0.75 and yields a pole-ratio of 3.5. Where this combination lacks in power density, it is predicted to make up for in cogging torque; it has an LCM of 252 compared to 132 of the integral slot case. Table 4-2 summarises the choices for the two winding configurations.

Table 4-2: Summary of pole and slot selection

	Integral Slot configuration	**Fractional Slot configuration**
Winding pole-pairs (p_s)	1	4
Rotor pole-pairs (p_r)	11	14
Pole ratio (G_r)	11	3.5
Slots (Z)	12	18
Slots / pole / phase (q)	2	0.75
Cogging factor (f_c)	2	2
LCM	132	252
Frequency (f) [Hz]	45.83	58.33

4.3 Winding Configuration

The spoke-type topologies employ circumferentially magnetised permanent magnets separated by steel pieces to provide a flux path as seen in Figure 4-1a and Figure 4-2a. The surface-mounted topologies consist of radially magnetised permanent magnets mounted on the surface of the rotor yoke as seen in Figure 4-1b and Figure 4-2b. The fractional slot case, shown in Figure 4-1, has 18 slots and 14 rotor pole-pairs, and thus by the Vernier principle, it is wound with 4 winding pole-pairs.

Figure 4-1: Fractional Slot Machine Configurations (a) FSST (b) FSSM

Similarly, for the integral slot case, shown in Figure 4-2, a 12 slot stator is wound with one winding pole-pair employing 11 rotor pole-pairs.

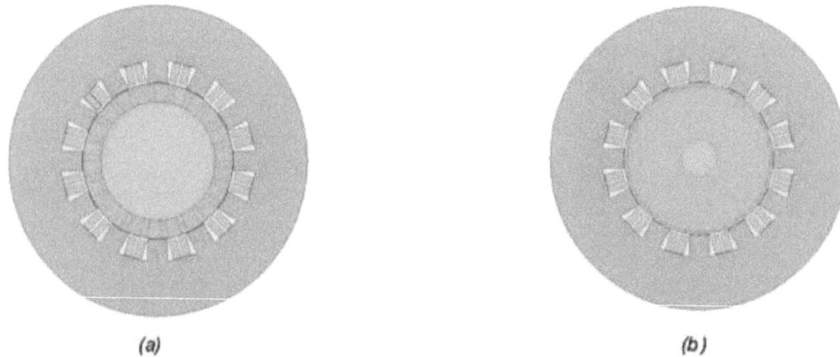

Figure 4-2: Integral Slot Machine Configurations (a) ISST (b) ISSM

All four machines adopt the open-slot structure, using the stator teeth to perform the flux modulation. The lack of tapered teeth maintains the air-gap variable permeance thereby enhancing the flux modulation effect. Double-layer windings are employed to ensure efficient use of stator and provide acceptable power density. Double-layer windings are also necessary for the short-pitched fractional-slot configuration. The winding configurations are shown in Figure 4-6.

Following the method outlined in Section 3.4.1, the star-of-slot diagrams were generated as follows:

1) The machine periodicity is determined using equation (3.4.2). For the fractional slot case, 18 slots and 4 winding pole-pairs, the periodicity is 2. For the integral slot case, 12 slots and 1 winding pole-pair, the periodicity is 1. This indicates that the fractional slot case possesses winding symmetry;
2) The number of spokes for the fractional slot case is $\frac{Z}{t} = 9$. Similarly, the integral slot case has 12 spokes;
3) The layout of the star of slots plane is determined by noting that these are 3-phase machines. Thus, the plane of both machines will consist of 6 sectors, each sector spanning 60^0. The result is shown in Figure 4-3;

Figure 4-3: The Star of slots sectors

4) Using equations (3.4.3) - (3.4.6), the angle between spokes is determined. For the fractional slot case it is 40^0. The angle between spokes for the integral slot case is 30^0;
5) The coil throw, given by equation (3.4.10), of the fractional slot case is 2. This implies that the coil pitch is given by $2 \times 40^0 = 80^0$. Similarly, the integral slot case has a coil throw of 6 and thus has a coil pitch of 180^0;
    a. For example, consider the fractional slot case. The first spoke is placed at 0^0 in sector A+. The second spoke resides 80^0 away in sector C-. And so the first 9 spokes are located.
    b. The remaining spokes are placed considering the periodicity of 2 for the fractional slot configuration as seen in Figure 4-4.

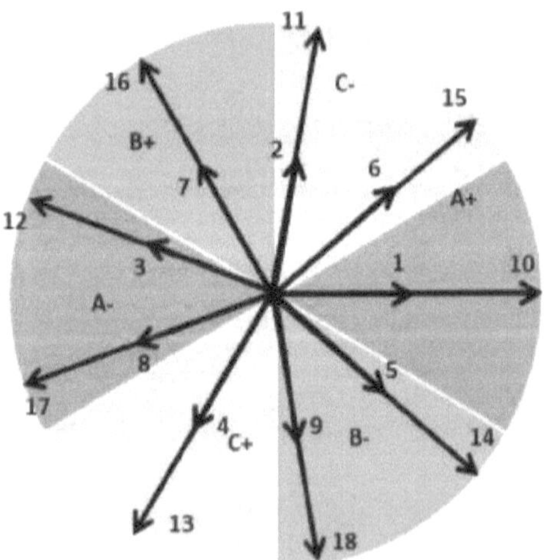

Figure 4-4: Fractional slot star-of-slot diagram

    c. Similarly, the integral slot star of slots diagram is generated as shown here in Figure 4-5.

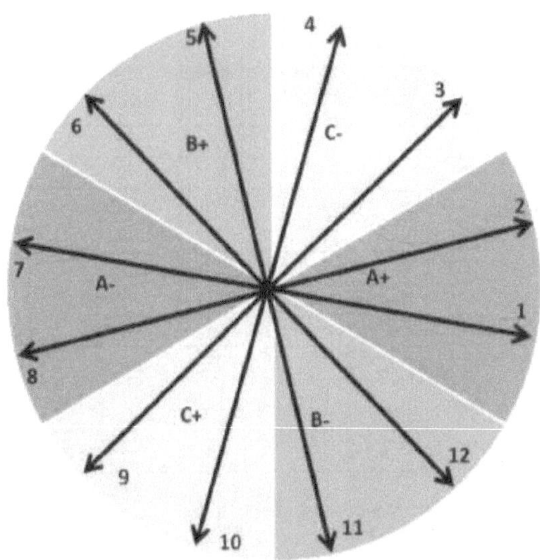

Figure 4-5: Integral slot configuration star-of-slot diagram

6) The star of slot diagrams are thus fully defined. The fractional slot case has 9 spokes, each displaced by 40^0. An odd number of spokes means that the number of spokes in the positive and negative sectors of a phase differ by 1. Spokes, or coil-ends, of the same coil are displaced by 80^0;
7) The arrows denote the spokes which represent the back-EMF generated in a single coil side. Considering that a coil-side of a phase winding resides inside a slot, the numbers on the diagram can represent a specific slot. To this end, the numbers indicate which slot houses which coil-side of a specific phase;
8) The star of slots diagram can then be used to complete the winding table as shown in Table 4-3 and Table 4-4. In the tables, "T" denotes coils which are situated at the top of the slot and "B" denotes coils at the bottom of the slot. The integral slot case is shown in Table 4-3.

Table 4-3: Integral slot coil locations

Coil Number	Phase	+ Slot	- Slot
Coil_1	A	3T	9B
Coil_2	A	4T	10B
Coil_3	A	3B	9T
Coil_4	A	4B	10T
Coil_5	B	11T	5B
Coil_6	B	12T	6B
Coil_7	B	11B	5T
Coil_8	B	12B	6T
Coil_9	C	7T	1B
Coil_10	C	8T	2B
Coil_11	C	7B	1T
Coil_12	C	8B	2T

The fractional slot case is shown in Table 4-4 below.

Table 4-4: Fraction slot coil locations

Coil Number	Phase	+ Slot	- Slot
Coil_1	A	6B	8T
Coil_2	A	10T	8B
Coil_3	A	10B	12T
Coil_4	A	15B	17T
Coil_5	A	1T	17B
Coil_6	A	1B	3T
Coil_7	B	7T	5B
Coil_8	B	3B	5T
Coil_9	B	16B	18T
Coil_10	B	16T	14B
Coil_11	B	12B	14T
Coil_12	B	7B	9T
Coil_13	C	9B	11T
Coil_14	C	13T	11B
Coil_15	C	13B	15T
Coil_16	C	18B	2T
Coil_17	C	4T	2B
Coil_18	C	4B	6T

9) The data in Table 4-3 and Table 4-4 is entered into ANSYS RMxprt which then generates the winding configuration as shown in Figure 4-6.

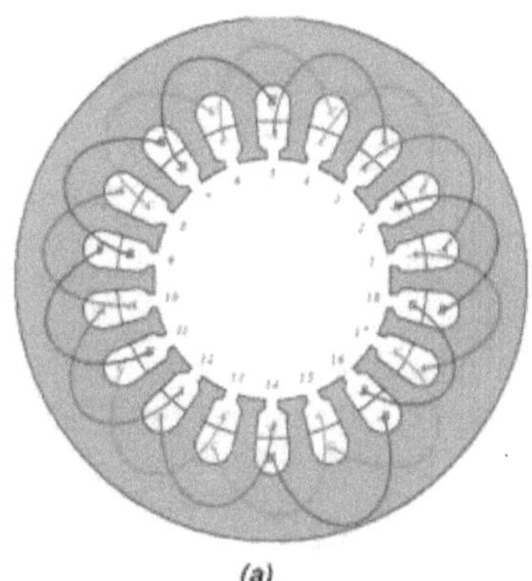

(a)

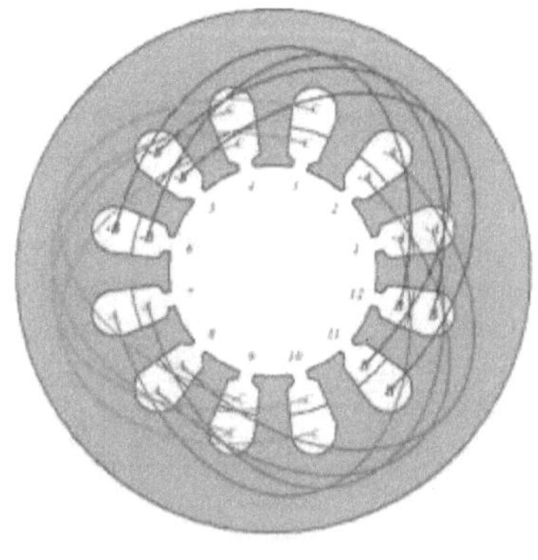

(b)

Figure 4-6: Winding layout (a) Fractional slot case (b) Integral slot case

Table 4-5 below indicates the parameters used to generate the winding layout for the fractional and integral slot winding configurations.

Table 4-5: Winding configuration layout defining parameters

	Integral Slot configuration	Fractional Slot configuration
Winding pole-pairs (p_s)	1	4
Slots (Z)	12	18
Slots / pole / phase (q)	2	0.75
Coils / phase	4	6
1st Harmonic Winding Factor (k_{w1})	0.966	0.945
3rd Harmonic Winding Factor (k_{w3})	0.71	0.58
Machine periodicity (t)	1	2
Spokes per phase (q_{ph})	4	3
Coil span / winding pole pitch (y_q)	6	2.25
Angle between spokes (α_{ph}) [$mech^0$]	80	180

4.4 Magnet Grade

As was explained through Betz's Limit, a wind turbine ideally has high efficiency and high power density. In electrical machine design, these requirements can be fulfilled through the implementation of rare-earth permanent magnets. These magnets offer high remanent flux and coercivity; providing protection against demagnetization. Unfortunately, high energy permanent magnets are costly [18]. Therefore, the designs in this study attempt to minimise cost by selecting rare earth permanent magnets with relatively small thicknesses, thereby saving on cost/PM volume but reserving energy density.

Indeed, it is the Vernier effect which compensates for this lack of PM volume. In fact, an interesting deduction that is made from the analysis of VPM machines is the inversely proportional relationship between average torque and PM thickness. In conventional motors, you would intuitively expect thicker magnets to increase average torque. Indeed, this is evident in VPM analysis if you consider the constant permeance component in equation (3.1.3) only. In fact, it is the first harmonic permeance component (equation (3.1.4)) that is inversely proportional to PM thickness, and moreover; thicker permanent magnets yield a lower Vernier contribution to average torque [15] [14]. Generally, PM thicknesses in VPM machines are chosen from the range 2-3mm. Such small PM thicknesses still offers appreciable torque density as well as low susceptibility to PM irreversible demagnetisation [79]. Thus, for this prototype, the initial PM thickness is chosen to be 3mm. The selected magnet grade is Neodymium Iron Boron (NdFeB) which has a remanence of 1.23T and a coercivity of 890kA/m.

The pole-arc to pole-pitch ratio of the surface mounted cases are designed to be 0.9. This was determined using the optimal pole-arc ratio to mitigate cogging torque [68]. The optimal pole-arc is given by:

$$\alpha_p = \frac{N - k_1}{N} + k_2 \tag{4.4.1}$$

Where $N = \frac{LCM\{Z,2p_r\}}{2p_r}$ and $k_1 = 1,2,..,N-1$. The pole-to-slot flux leakage is compensated for by k_2 which is usually in the range: 0.01-0.03.

Table 4-6: Initial permanent magnet specifications

	Integral Slot configuration	Fractional Slot configuration
Rotor pole-pairs (p_r)	11	14
Material	NdFeB	
Remanence (B_r) [T]	1.23	
Coercivity (H_c) [kA/m]	890	
Relative Permeability (μ_r)	1.099	
Thickness (l_m) [mm]	3	
Pole-arc (α_p)	0.9	

The spoke type PM parameters are omitted here as those will be sized in the FEA phase

4.5 Air-gap Diameter

Essentially, VPM machines make use of the fundamental flux density component and a slot harmonic component termed the 'Vernier contribution'. The back-EMF is thus a sum of two terms induced by these two flux density components. To formulate the VPM power equation, the Vernier back-EMF and machine electric loading (A_s) is taken account of. Following on from [80], [10] the back-EMF is given by:

$$E_{0m} = k_{leak} k_w N_{ph} (B_0 + G_r B_1) \omega_r L_{st} D_g \tag{4.5.1}$$

Where: k_{leak} is the flux leakage factor, B_0 is the amplitude of the air-gap fundamental flux density, B_1 is the amplitude of air-gap effective harmonic flux density, and G_r is the 'gearing ratio' which is equal to $\frac{p_r}{p_s}$. The peak phase current can be written in terms of the electric loading A_s:

$$I_{am} = \frac{\sqrt{2}A_s\pi D_g}{2mN_{ph}} \quad (4.5.2)$$

Then using the well-known power relationship between voltage and current in an m-phase machine

$$P_0 = \frac{m}{2}E_{0m}I_{am}\cos(\psi) \quad (4.5.3)$$

The obtainable power capacity of a VPM generator is expressed as:

$$P_0 = \frac{\sqrt{2}\pi}{4}k_{leak}k_w(B_0 + G_rB_1)A_s\omega_rL_{st}D_g^2\cos(\psi) \quad (4.5.4)$$

Where ψ is the angle between the no-load back-EMF and phase current, and $\cos(\psi)$ represents the inner power factor. This equation was used to size the air-gap diameters of the four machines.

$$D_g^2 = \frac{P_0}{\frac{\sqrt{2}\pi}{4}k_{leak}k_w(B_0 + G_rB_1)A_s\omega_rAR\cos(\psi)} \quad (4.5.5)$$

AR is the aspect ratio. For wind applications, the generators are usually designed with large outer diameters. This would not be problematic in terms of centrifugal forces as the machine operates at low speeds. The aspect ratio was chosen to be 0.95 in this study.

4.6 Magnetic Loading and Magnetic Circuit Geometry

The flux density components are determined from the fundamental component:

$$B_0 = F_1P_0 \quad (4.6.1)$$

And the slot harmonic (Vernier) component:

$$B_1 = \frac{F_1P_1}{2} \quad (4.6.2)$$

Where the fundamental PM MMF is

$$F_1 = \frac{4}{\pi}\frac{B_r}{\mu_r\mu_0}h_m\sin\left(\alpha_p\frac{\pi}{2}\right) \quad (4.6.3)$$

The permeance expressions, equations (3.1.3) and (3.1.4), are functions of the slot geometry; namely the parameters $c_0 = \frac{b_0}{\tau}$ the ratio of slot opening to slot pitch, and $\frac{b_0}{g'}$ the ratio of slot opening to magnetic

air-gap length. The average slot pitch is given as $\tau = \frac{\pi D_g}{Z}$. The slot and teeth widths are readily computed using the calculated air-gap diameter and by choosing appropriate values for $\frac{b_0}{\tau}$ and $\frac{b_0}{g'}$.
The Vernier magnetic loading $\left(B_0 + \frac{p_r}{p_s} B_1\right)$ is thus fully defined by the slot, stator and PM geometry.

Parallel, un-tapered teeth are chosen so that the Vernier effect is maintained by guaranteeing a variable air-gap permeance. As explained earlier, the Vernier effect is utilised most effectively by selecting $c_0 = 0.5$, such that the slot and teeth widths are equal. To avoid saturation in the teeth, the tooth width is thus constrained by:

$$w_{tb} = \frac{\pi D_g B_0}{Z K_{st} B_{sat}} \tag{4.6.4}$$

The magnetic loading and air-gap diameter is used to compute the stator yoke thickness [81]:

$$w_{sy} = \frac{\pi D_g B_0}{4 p_s K_{st} B_{sat}} \tag{4.6.5}$$

And the rotor thickness:

$$w_{ry} = \frac{\pi D_g B_0}{4 p_s K_{st} B_{sat}} \tag{4.6.6}$$

B_{sat} is the saturation flux density of the ferromagnetic material and K_{st} is the lamination stacking factor. These equations show that the rotor and stator yoke widths are inversely proportional to the number of winding pole-pairs.
The saturation flux density is obtained from the datasheet for the selected electrical steel. M530-50a non-orientated laminated electrical steel was selected. The B-H curve for M530-50A is shown in Figure 4-7. The saturation flux density is taken from the knee-point of the B-H curve.

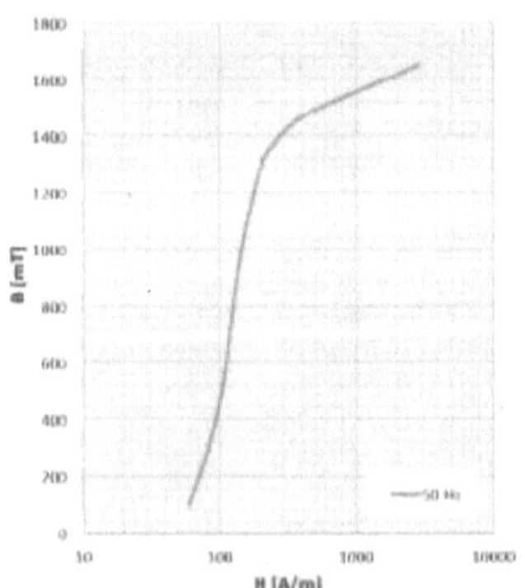

Figure 4-7: M530-50A datasheet B-H curve [82]

A cost-effective choice, M530-50A exhibits minimal core loss per kilogram at low frequencies. This being a low frequency machine, a lower core loss steel was unnecessary as this would incur additional avoidable cost in a real-world design.

Table 4-7: Magnetic Circuit parameter summary

	Integral Slot configuration	Fractional Slot configuration
Lamination material	M530-50A	
Lamination thickness [mm]	0.5	
Lamination stacking factor	0.95	
Lamination saturation flux density (B_{sat}) [T]	1.4	
Air-gap Diameter (D_g) [mm]	166.23	208.95
Stack length (L_{st}) [mm]	159.02	199.49
Aspect Ratio (AR)	0.95	0.95
Fundamental flux density (B_0) [T]	0.85	0.89
Vernier harmonic flux density (B_1) [T]	0.21	0.2
Tooth width (w_{tb}) [mm]	21.86	18.05
Stator yoke thickness (w_{sy}) [mm]	59.2	20.06
Rotor thickness (w_{ry}) [mm]	80.12	19.61
Slot opening to slot pitch ratio (c_0)	0.5	
Slot pitch (τ) [mm]	43.94	36.75
Slot opening to magnetic air-gap length ratio (b_0/g'')	6.17	5.3
Air-gap length (l_g) [mm]	0.8	

4.7 Slot Geometry and Electric Loading

To facilitate comparisons among the four designs, the four machines share the following characteristics:

- Slot-fill factor
- Current density

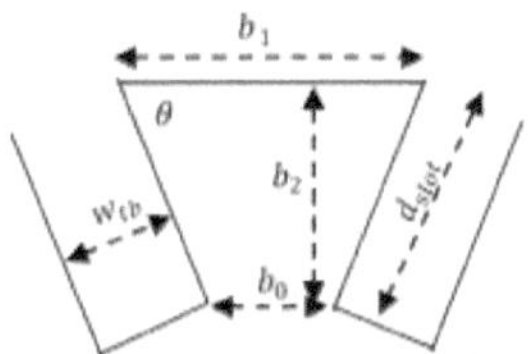

Figure 4-8: Slot geometry used to size stator slots and teeth

As a result of the parallel teeth, the slots are trapezoidal; the geometry is shown in Figure 4-8. And the slot area is given by

$$A_{slot} = \frac{b_1 + b_0}{2} \times b_2 \tag{4.7.1}$$

Additionally, the slot area can be described by the electrical parameters:

$$A_{slot} = \frac{N_{cs} A_{cond}}{K_{fill}} \tag{4.7.2}$$

N_{cs} is the number of turns per slot, A_{cond} is the area of the conductor and K_{fill} is the slot-fill factor.

The area of the conductor is determined from the current density and rated current specification of 10A.

$$A_{cond} = \frac{I_{RMS}}{J_{RMS}} \tag{4.7.3}$$

A design choice, the current density was chosen to be 3.8 $A.mm^{-2}$. The chosen current density depends on the cooling system that is implemented and the power rating of the machine. In this study, the designs are totally enclosed without an external cooling system and the power ratings are in the range 8-10 HP. Therefore, a relatively low current density is chosen as per Table 4-8.

Table 4-8: Recommended current density per machine cooling technique [83]

Condition	**A/mm²**
Totally enclosed	1-5
Air-over fan-cooled	5-10
Liquid cooled	10-30

Additionally, by equation (4.7.3) for a given current, it would be desirable to have a low current density to maximise the conductor area. A large conductor area minimises winding resistance and improves the thermal conduction of heat from the conductors to the lamination stack [83].

By choosing the slot-fill factor and current density to be constant across the four designs, the slot current density, and thus the I^2R losses per (unit slot volume) slot are constant. This places each of the designs on equal footing in terms of thermal capabilities. The resistance per slot is:

$$R_{slot} = \frac{\rho L_{st} 4 N_{tc}^2}{K_{fill} A_{slot}} \tag{4.7.4}$$

Using equation (4.7.2)

$$R_{slot} = \frac{\rho L_{st} 4 N_{tc}^2}{N_{cs} A_{cond}} = \frac{\rho L_{st}}{A_{cond}}(2N_{tc}) \tag{4.7.5}$$

$$= \frac{J_{RMS}}{I_{RMS}} \rho L_{st}(2N_{tc})$$

Note the inverse proportionality of conductor area to slot resistance. The above holds true for machines having the same stack length and number of turns/coil. With this in mind, the slot area is well defined by equations (4.7.1) and (4.7.2) and the relevant design choices explained. Hence, by using basic trigonometry and geometry in Figure 4-8, the angle between the tooth base and stator yoke is $\theta = \frac{\pi(Z-2)}{2Z}$, and therefore the slot depth can be determined from solving the following quadratic for d_{slot}:

$$A_{slot} = (d_{slot}\cos(\theta) + b_0) \times d_{slot}\sin(\theta) \tag{4.7.6}$$

With the electric and magnetic loadings fully defined, winding performance characteristics such as resistance can be calculated. The electric loading parameter summary is shown in Table 4-9.

Table 4-9: Electric Loading parameter summary

	Integral Slot configuration	Fractional Slot configuration
Slot Depth (d_{slot}) [mm]	22.52	16.01
Slot Opening (b_0) [mm]	21.76	18.67
Slot area (A_{slot}) [mm²]	620.34	338.28
Turns per coil (N_{tc})	36	30
Turns per coil per phase (N_{ph})	144	180
Conductor area (A_{cond}) [mm²]	2.89	2.63
Slot fill factor (K_{fill})	0.75	0.75
Current density (J) [A/mm²]	3.8	3.8
Per phase resistance (R_{ph})[Ω]	1.38	1.26
Air-gap Inductance [mH]	106.46	23.24
Slot leakage Inductance [mH]	1.75	2.16
End turn leakage Inductance [mH]	0.76	0.13

5. FEA Comparative Study

5.1 No-load Analysis

5.1.1 No-load Flux Density

The verification of the Vernier effect is completed by analysing the air-gap flux density plots in Figure 5-1 and Figure 5-2. The flux density distributions reveal the presence of the p_s winding pole-pairs present in the stator field distribution.

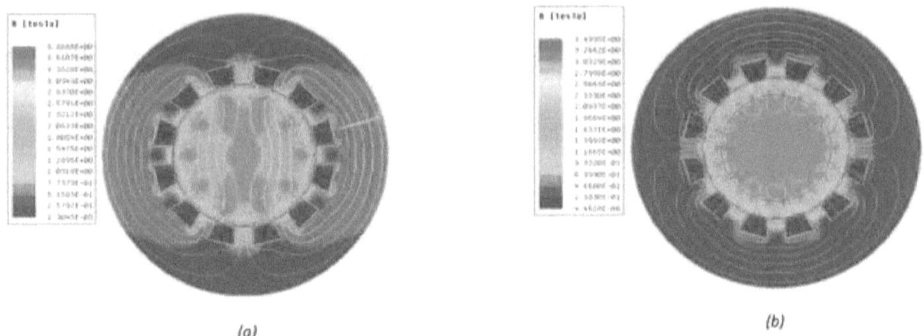

(a) (b)

Figure 5-1: No-load flux density distributions (a) ISSM (b) ISST

The 11 rotor pole-pair in the integral slot case produced 1 winding pole-pair as can be seen in Figure 5-1a and Figure 5-1b. Similarly, for the fractional slot case, in Figure 5-2a and Figure 5-2b, the 14 rotor pole-pairs produced 4 winding pole-pairs. Evidently the stators in both fractional and integral slot cases performed modulation of the rotor flux density.

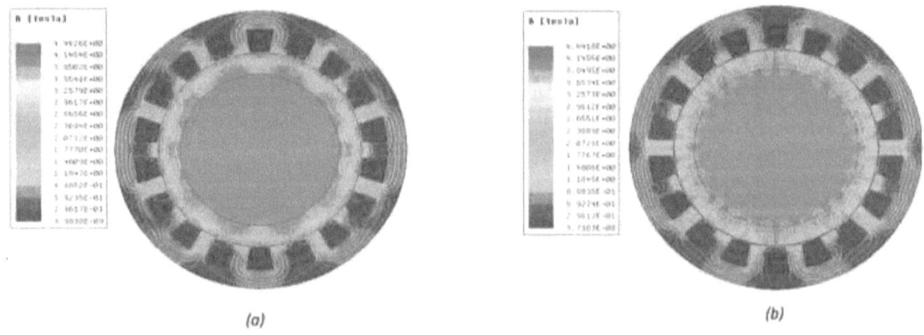

(a) (b)

Figure 5-2: No-load flux density distributions (a) FSSM (b) FSST

The flux density harmonic spectra provides further verification of the Vernier effect by indicating the presence of the fundamental harmonic , p_r , and two slot harmonics, $(Z - p_r)$ and $(Z + p_r)$, in accordance with equation (3.1.7). The integral slot flux density harmonic spectra is shown in Figure 5-3.

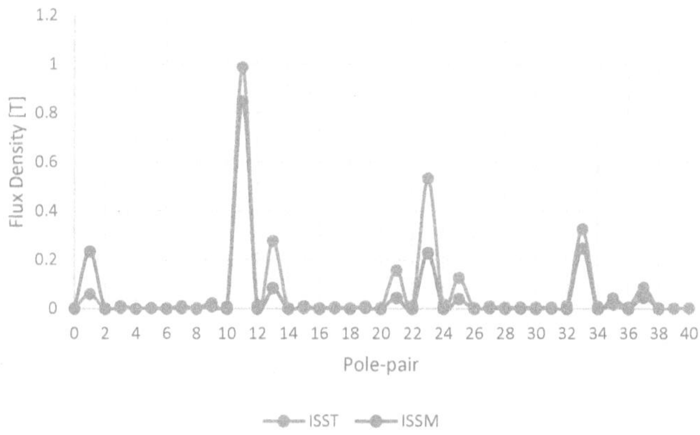

Figure 5-3: Flux density harmonic spectrum for the integral slot case

A comparison of the spoke and surface mounted harmonic spectra in Figure 5-4 shows that the spoke fundamental flux density harmonic is higher than that of the SM cases due to the spoke configurations' flux focusing feature.

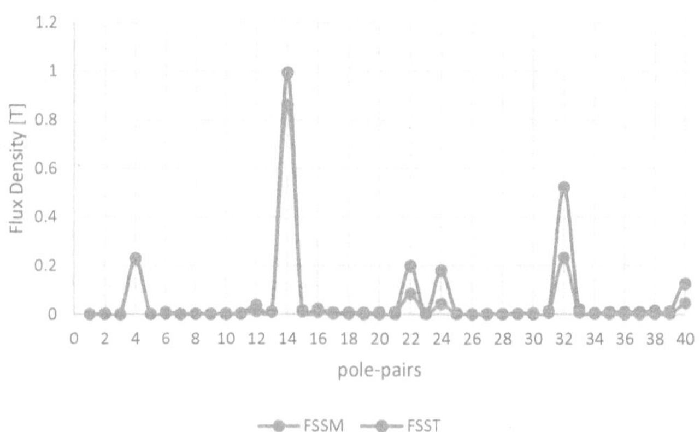

Figure 5-4: Flux density harmonic spectrum for the fractional slot case

The fundamental peak flux density for the ISSM is 0.8493T, whilst the ISST produces 0.9897T. For the fractional slot case, the FSSM case has a peak of 0.8600T and the FSST a peak of 0.9933T. Divergence from the Vernier principle is noted in the ISST case in Figure 5-3; the 1 pole-pair slot harmonic is clearly suppressed. Equation (3.1.7) indicates that the slot harmonics should have the same magnitude. This is true in the ISSM, FSSM, FSST but not the ISST case. This is clear evidence that the Vernier effect is suppressed in the integral slot spoke type topology, as postulated in [23]. The flux density waveforms are shown in Figure 5-5.

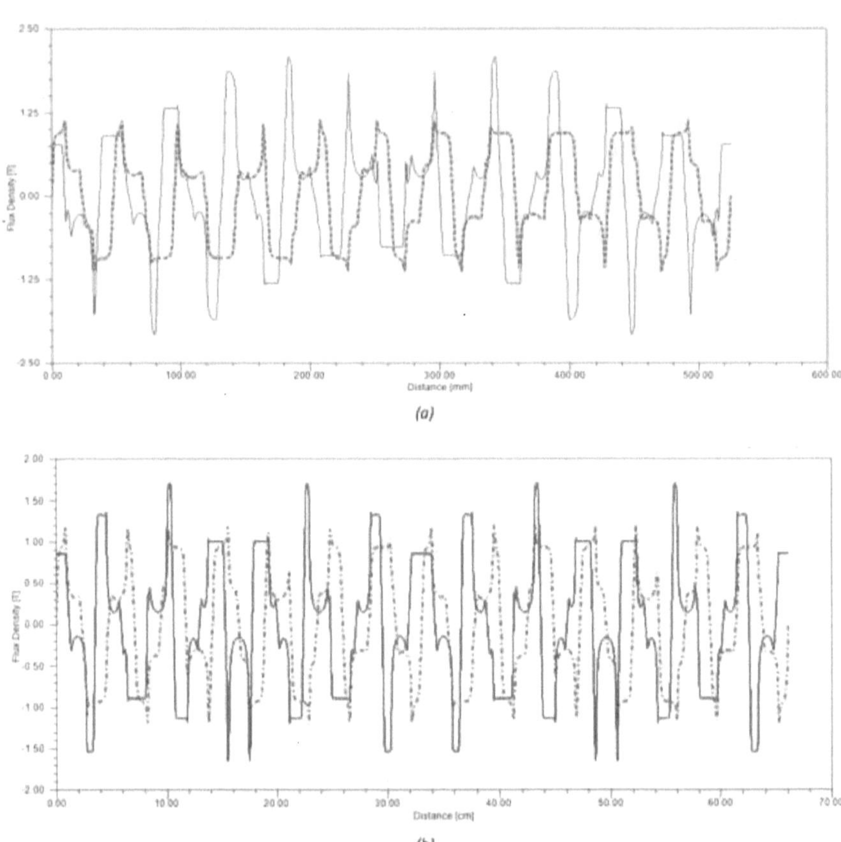

Figure 5-5: No-load flux density waveforms (a) integral slot case (b) fractional slot case

Continuing with no-load tests, the machines were simulated at rated speed of 250RPM under open-circuit conditions.

5.1.2 Open Circuit Back-EMF

For the surface-mounted designs, the literature is well defined enough to entrust the calculated PM thicknesses to yield the desired back-EMF. However, the literature for the spoke designs is not conclusive enough. Hence, the magnet thickness is parameterised and varied to determine the thickness which will yield the desired back-EMF. The PM thickness is varied from 2mm to 6mm in accordance with [79]. The analysis of the FSST case, as shown in Figure 5-6, reveals that the targeted phase back-EMF of 225V can be achieved using 30 turns per coil. The same as the FSSM case, albeit by increasing the PM thickness to 4mm.

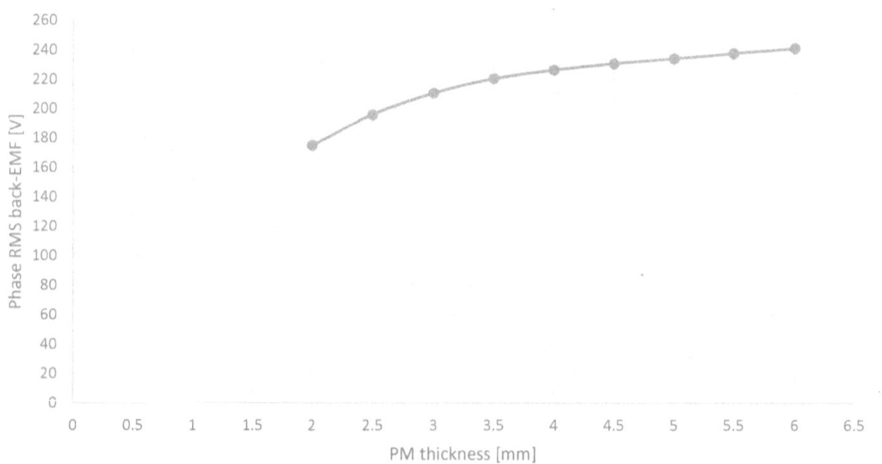

Figure 5-6: Graph depicting back-EMF variation with respect to PM thickness for the FSST case

However, for the ISST case the targeted back-EMF cannot be achieved at all by using the same turns per coil as the ISSM case. Figure 5-7 reveals that using 30 turns per coil will achieve a back-EMF of 80V RMS using 6mm thick permanent magnets. Thus, the turns per coil was chosen to be 80 and the PM thickness 5mm.

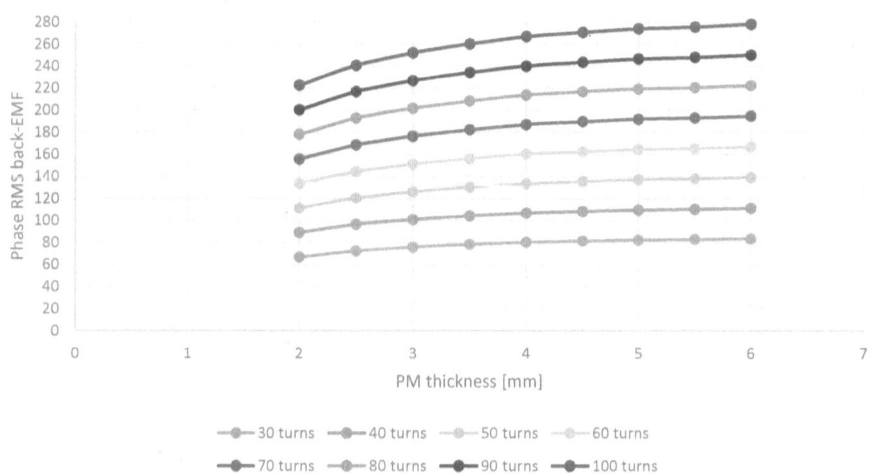

Figure 5-7: Graph depicting back-EMF variation with respect to PM thickness and turns per coil for the ISST case

The back-EMF characteristics are compared in Figure 5-8. As explained in the Sizing section, the four machines were designed to generate the same back-EMF magnitude of 380V line-line.

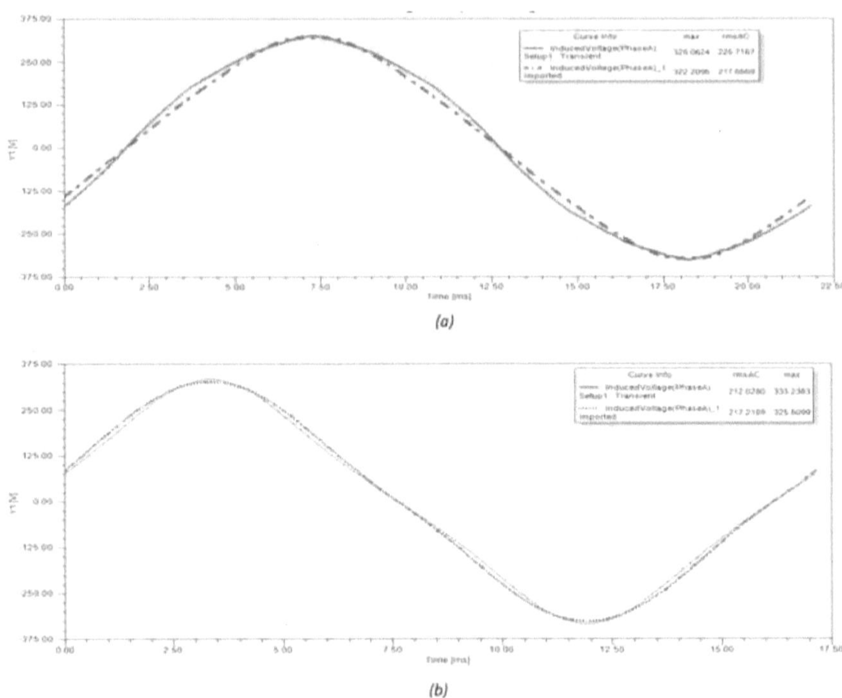

Figure 5-8: No-load back-EMF waveforms (a) integral slot case (b) fractional slot case

As a result, there are variations in the four designs as seen in Table 5-1. Ideally, of the four machines, the chosen machine would produce the desired back-EMF using the fewest number of turns/coil, smallest PM volume and having the smallest rotor volume.

Table 5-1: Design parameters for back-EMF

	Air-gap diameter [m]	**Turns/coil**	**PM volume [cm^3]**
ISSM	166.23	36	219.76
FSSM	208.95	34	351.95
ISST	166.23	80	381.65
FSST	208.95	30	458.03

From Table 5-1, the integral slot has the smallest air-gap diameter due to its higher pole-ratio and by equation (4.5.5). The turns/coil of the machines are more-or-less in the same range (bar a few slight inconsistencies). However, the ISST case requires 80 turns/coil to generate the same voltage – more than double that of any of the other three cases. Moreover, the ISST uses the largest amount of PM and winding material. Looking at the harmonic spectra, the spoke configurations produced the highest flux density amplitudes, due to their flux focusing feature. But closer inspection reveals that the synchronising slot harmonic $(Z - p_r)$ is significantly lower than in the surface-mounted configuration.

The suppression of the Vernier effect is believed to be the reason for the diminished power density in the ISST case. Interestingly though, this is not the case for the FSST configuration. The slot harmonic in the fractional slot configurations are relatively similar in magnitude. Harmonic analysis confirmed that the machines with the highest pole-ratio produced the lowest THD; 4.72% for the ISST case and 2.43% for the ISSM case, as was predicted in [16].

Figure 5-9: back EMF harmonic spectrum for the integral slot designs

Compared to the lower pole-ratio fractional slot machines produced the highest THD; 6.84% for the SM case and 10.64% for the spoke case. Notably, the fractional slot spoke case produced more than double the THD of the integral slot spoke THD. The high THD of the FSST case can be expected due to the large third harmonic. Without short-pitching the FSST windings, the 3rd harmonic winding factor remains unopposed.

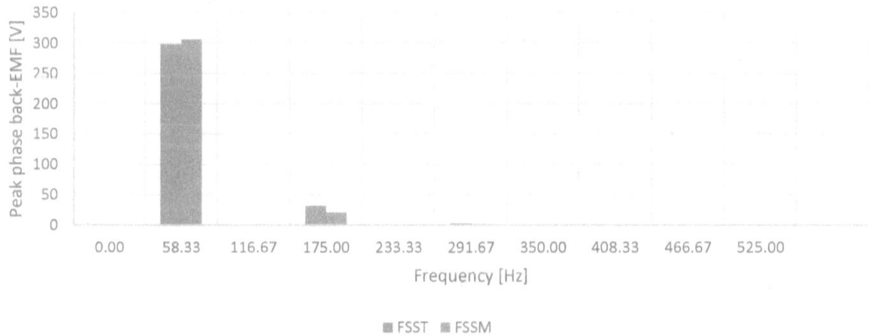

Figure 5-10: EMF harmonic spectrum for the fractional slot designs

5.1.3 Cogging Torque

Figure 5-11a gives the cogging torque of the integral slot spoke and SM cases. The two machines share the same cogging period as they have the same pole-pair and slot combinations. The peak-peak cogging of the SM case is 2.35Nm compared to the spoke case 1.03Nm.

Figure 5-11b gives the cogging torque of the fractional slot spoke and SM cases. The peak-peak cogging of the SM case is 3.16Nm compared to the spoke case 9.05Nm. The spoke type introduces a salient effect which distorts the no-load flux density and exacerbates cogging torque [84]. This is evident in the fractional slot case, but not so in the integral slot spoke case.

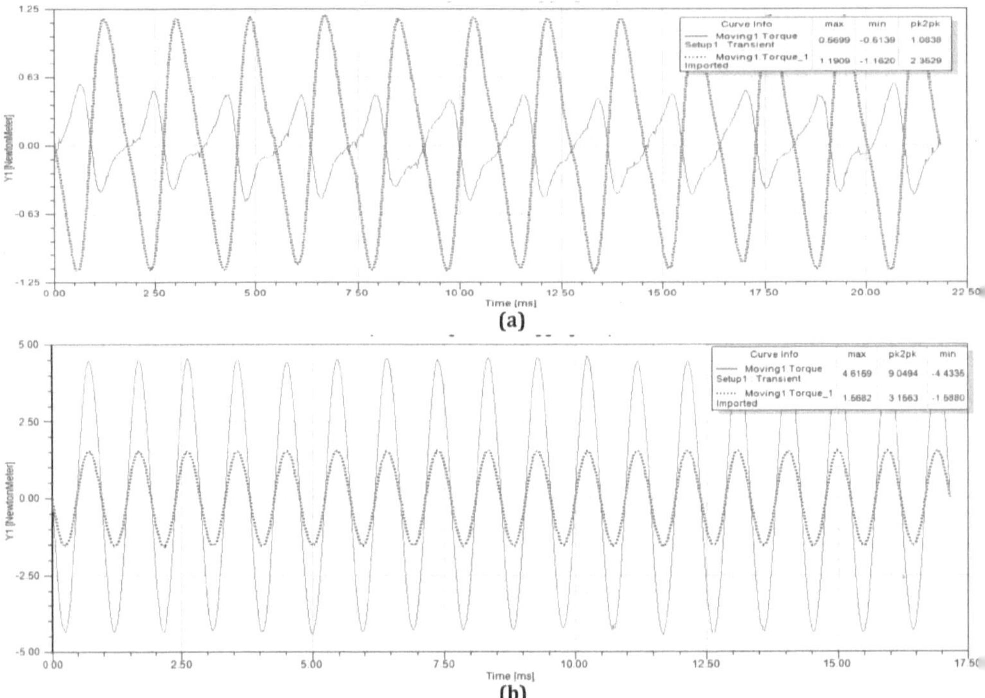

Figure 5-11: Cogging torque waveforms. Spoke (red) SM (blue). (a) Integral slot case (b) fractional slot

Closer inspection shows that the flux per pole of the ISST case is significantly less than the other cases as seen in Table 5-2. This explains its need for double the amount of turns/coil to meet the back-EMF requirement.

Table 5-2: Peak flux no-load flux linkage and flux

	ISSM	ISST	FSSM	FSST
Flux linkage [Wb]	1.0518	1.1180	0.5914	0.5756
No-Load phase Flux [Wb]	0.0292	0.0035	0.0174	0.0192

Thus, by equation (3.7.5), the ISST case would produce the lowest cogging torque due to its significantly lower flux per pole – it is 10 times less than the SM case. This is acknowledged in [32] where the authors claim that the low cogging torque in Vernier machines owes to the large amount of leakage flux. A high leakage flux reduces the redundant magnetic field harmonics thereby reducing redundant EMF harmonics as well, which in turn reduces cogging torque.

5.2 Full-load Analysis

5.2.1 Full-load Torque

The machines were simulated in motor mode. Balanced three-phase currents were injected into the windings. The d-axis current was set to zero. As described earlier, the machines were designed to produce the same nominal torque of 230Nm. The FEA results show that this was accomplished. The torque FEA plots reveal the different levels of torque ripple amongst the four designs. The results of the torque study are shown in Table 5-3.

Table 5-3: Torque Characteristics

	Torque [Nm]	Torque/rotor volume [kNm/m^3]	Rated current [A]	PM thickness [mm]	% Torque ripple
ISSM	236.15	67.92	11	3	0.97
FSSM	233.06	34.15	10	3	2.36
ISST	223.55	66.76	11	5.75	4.45
FSST	228.24	33.5	10	4	5.05

The integral slot machines offer the greatest torque/rotor volume. This is expected as they have the higher pole-ratio. However, the ISST case required thicker permanent magnets despite its high pole-ratio and larger no-load peak flux density – almost double the PM thickness of the SM case. The natural conclusion is that the majority of the ISST flux is not going into power conversion and is rather lost in the form of leakage flux. This is also apparent in the flux density plots of Figure 5-1. All four machines have ripple percentages less than 10% which is desirable in mitigating turbine vibrations. Notably, the ISST case, which yielded a cogging percentage of 0.45%, produces a large amount of torque ripple at 9.93Nm peak-peak. As torque ripple is the result of cogging and parasitic flux density harmonics, it is safe to conclude that it is its rich harmonic spectrum which is the cause of the high ripple. High cogging torque and by extent torque ripple is a fundamental characteristic of the spoke-type topology.

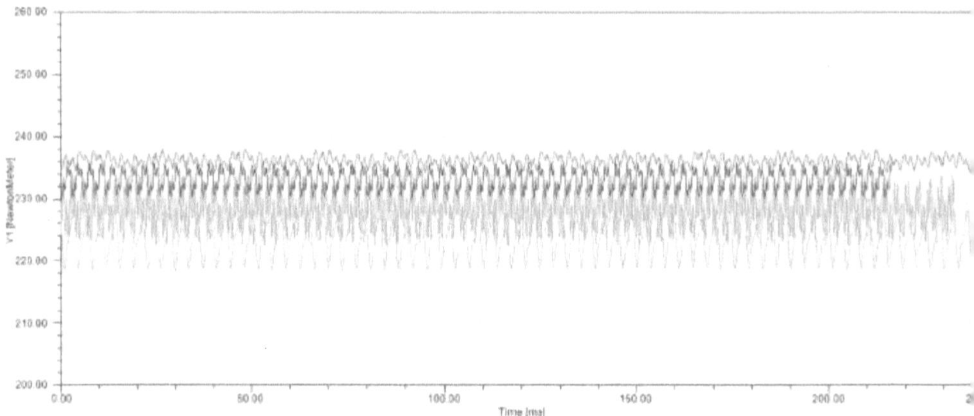

Figure 5-12: Torque FEA waveforms - ISSM (red), ISST (orange), FSSM (blue), FSST (green)

5.2.2 Inductance and Power Factor

By equation (3.5.11), the machines with the higher pole-ratio will exhibit the largest reactance and thus produce the lowest power factor. Accordingly, the synchronous inductances were studied. The integral slot machines have a larger winding pole pitch; 6 compared to 2 of the fractional slot case, resulting in longer end-turns and thus greater end-turn leakage inductance. Once again it is the ISST which falls short in terms of performance. It has the largest synchronous inductance which can be attributed to the large winding pole pitch as well as the previously noted leakage flux.

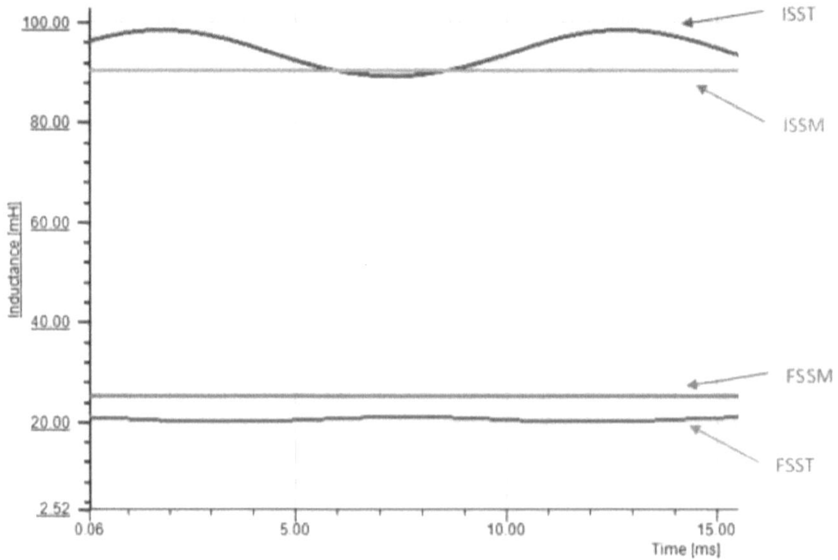

Figure 5-13: Synchronous Inductances

It was shown in equations (3.5.11) and (3.6.3) that it is the large synchronous reactance which is the predominating reason for poor factor in Vernier machines. Indeed, this notion is confirmed by the ISSM case, where the inductance is three times larger than the fractional slot cases, as shown in Figure 5-13, yielding the lowest power factor of the four cases.

What follows is an analysis of the machine power factor. To study the power factors, a sinusoidal current was injected into the windings and plotted on the same axes as the terminal voltage. The power factors were derived from the phase difference between the two waves. By equation (3.6.3), Vernier theory dictates that the higher pole-ratio VPMs would have lower power factors. This notion is confirmed in Figure 5-14 and Table 5-4.

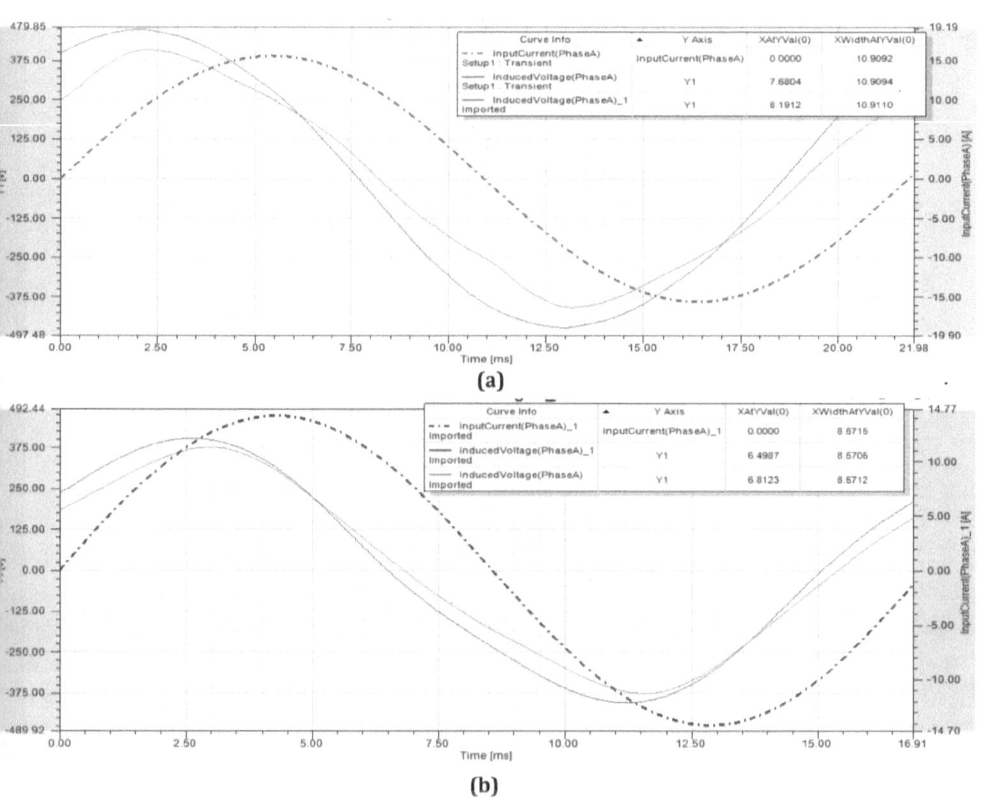

Figure 5-14: Power Factor Characteristics (a) integral slot case (b) fractional slot case

The ISSM case had the lowest power factor of the four machines. However, the spoke type configurations display the best power factor over-all; even though the ISST had a larger pole ratio, it still produced a similar power factor to that of the FSSM case.

Table 5-4: Power Factor

	Phase difference [ms]	PF1	PF2
ISSM	3.23	0.6	0.62
ISST	2.72	0.71	0.70
FSSM	2.07	0.72	0.71
FSSM	1.76	0.8	0.74

Moreover, the phase displacement method (PF1) is compared with the harmonic method from Section 3.6.3 (PF2). It is clear that when the THD is factored into the PF calculation, it has a significant effect.

5.2.3 Loss and Efficiency

The core losses and winding losses are analysed at full-load conditions. The same rated current of 10A RMS is injected into each of the four designs. The copper loss is thus determined by that same input current and the winding resistance. The core losses are determined from the manufacturer power vs frequency curves. The curves are easily obtained from the manufacturer datasheet [82]. The total losses and efficiencies of the four machines are compared in Figure 5-15 tabulated in Table 5-5. All four machines offer relatively good efficiency. Notably, the ISST case produces the lowest core loss which can be attributed to its high flux leakage and low flux linkage. Furthermore, the 80 turns per coil of the ISST design yielded the largest copper loss.

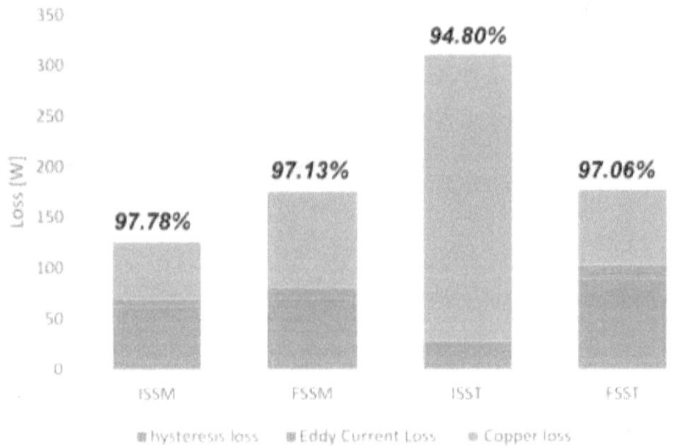

Figure 5-15: Full-load losses and efficiencies

5.2 FEA Comparative Study Conclusion

In this chapter, a comparative study consisting of four Vernier PM topologies; fractional slot spoke, fractional slot SM, integral slot spoke and integral slot SM; was completed. The four machines were explicitly analysed and compared in terms of flux density distribution, back-EMF, Torque characteristics, power factor and losses. These factors were used to determine the topology's suitability for a wind turbine application.

The spoke topology produced the largest magnetic loading, confirming the flux-focusing effect. The integral slot spoke-type topology was the only case where the Vernier effect was not prevalent, the synchronising slot harmonic was suppressed. The implication of this was larger material consumption was needed to meet the back-EMF and torque requirements – the integral slot spoke case required the greatest amount of copper wiring and PM material.

Fractional slot spoke produced the worst performance in terms of cogging and torque ripple. This was attributed to the salient effect of the spoke structure as well as the increased air-gap flux density. By Figure 5-2 the fractional slot spoke case displayed the least amount of leakage flux, whereas the integral slot spoke case displayed the largest amount. As expected, the integral slot SM produced the largest torque/unit rotor volume due its larger pole-ratio. However, the integral slot SM case produces the lowest power factor. Integral slot spoke type is the costliest of the four machines as it has the greatest material consumption. In terms of torque density, cogging and ripple, the integral slot SM case is the most attractive. However, it falls short in terms of power factor. Conversely, the fractional slot spoke type produces the largest power factor but falls short in terms of cogging torque and torque ripple.

The fractional slot SM case fulfils all the criteria required for wind applications but yields a lower power factor compared to the fractional slot spoke case. This can be attributed to larger leakage flux in the SM topology compared to the spoke. Therefore, providing cogging and ripple lower than 15% of nominal and adequately low material consumption, whilst providing the highest power factor; the fractional slot spoke case is deemed the most suitable for a wind power application. It is believed that the power factor and torque performance can be improved through optimisation.

Table 5-5: Comparative study outcomes

	ISSM		ISST		FSSM		FSST	
	Analytical	FEA	Analytical	FEA	Analytical	FEA	Analytical	FEA
Generated Power [W]	6000							
Torque magnitude [Nm]	-	236	-	224	-	233	-	228
Torque ripple [%]	-	0.97	-	4.45	-	2.36	-	5.05
Cogging torque [%]	-	1	-	0.46	-	1.36	-	3.97
Back-EMF [V]	225	226	-	217	217	216	-	211
Power Factor	0.6	0.62	-	0.70	0.69	0.71	-	0.74
Voltage THD [%]	-	2.43	-	4.72	-	6.84	-	10.64
Hysteresis Loss [W]	69.88	61.66	-	24.83	75.81	70.33	-	88.65
Eddy Current Loss [W]	7.95	7.92	-	2.73	9.74	10.48	-	14.32
Copper Loss [W]	378	55.96	-	283.28	414	94.21	-	73.43
Resistance [Ω]	1.26	1.13	-	3.01	1.38	1.4	-	1.24
Inductance [mH]	108.97	90.42	-	94.2	25.53	25.29	-	20.64
Fundamental flux density (B_0) [T]	0.85	0.85	-	0.99	0.89	0.86	-	0.99
Vernier harmonic flux density (B_1) [T]	0.21	0.23	-	0.06	0.2	0.23	-	0.23

Evidently, FEA results of the surface mounted designs displayed good correlation with analytical calculations. The spoke type designs, without a well-defined analytical model, displayed understandable and explainable performance behaviour. Owing to the fact that the FSST yielded respectable power factor and that other performance criteria were within 10%, it is chosen for further analysis.

6. Torque and Current Angle Study

6.1 Introduction

The following analysis constitutes a phasor study of the FSST and ISSM designs. The two were chosen as they fell at opposite ends of the comparative study spectrum. The ISSM case displayed the greater torque density and the poorest power factor. Whilst, the FSST displayed acceptable torque density and yielded the greatest power factor. The ultimate aim being to determine the machine terminal power factor by plotting the voltage, current and open-circuit back-EMF phasors in the DQ plane. FEA full-load machine simulations can be conducted in ANSYS Maxwell using two different methods – current and voltage excitation. Two machines are analysed in this regard; the fractional slot spoke-type (FSST) and integral slot surface mounted (ISSM) machines.

6.2 Current Excitation – Current Angle Study

The first method is current excitation, in which rated 3-phase purely sinusoidal current signals are applied to the machine windings. The signals are described by:

$$i_{exc}(t) = \sqrt{2}I_{rms}\sin\left(2\pi ft + \gamma + \theta_{phase}\right) \tag{6.2.1}$$

I_{rms} and f is the machine rated RMS voltage and frequency respectively. The current angle is γ and θ_{phase} is the phase difference between the three phases $(0, \frac{2\pi}{3}, \frac{4\pi}{3})$. The current angle is varied from 0° to 180°. The torque over this range of both machines is shown in Figure 6-1. At zero degrees, both machines displayed rated torque. At 90° the machines exhibited zero torque. This shows that a current angle of 0° produces positive q-axis current, whilst 90° produces demagnetizing d-axis current.

Figure 6-1: Torque vs current angle

From 90° to 180° the current is producing negative q-axis and d-axis current. The current phasor is thus in the third quadrant and the two machines are evidently in generator mode, hence the negative torque. The current angle is defined as the angle between the current phasor and the open circuit (OC) back-EMF phasor. The aforementioned analysis can be verified by comparing the current waveforms to that of the OC back-EMF and noting the phase difference, as shown in Figure 6-2.

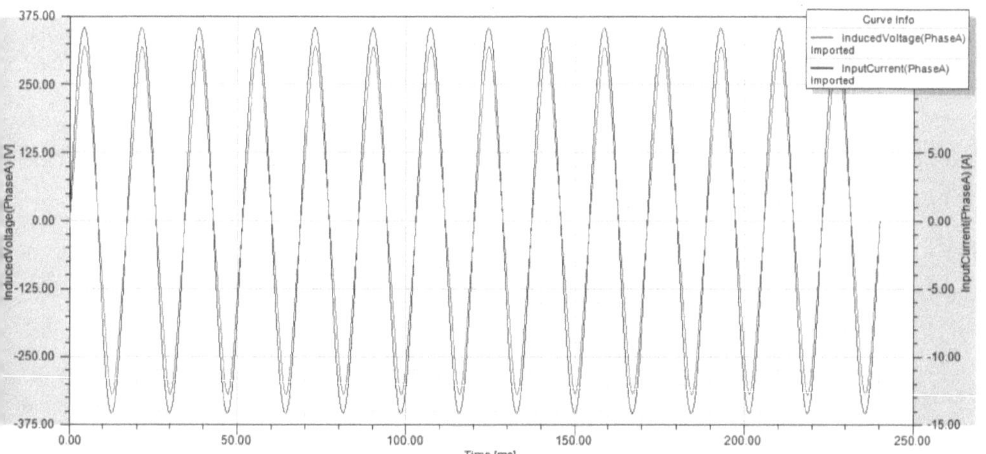

Figure 6-2: Current and back-EMF for γ = 0°

Figure 6-2 shows the OC back-EMF and current wave forms for a current angle of $\gamma = 0^o$. Clearly the two waveforms are in phase. Thus, when the current angle is zero, the current phasor lies along the positive q-axis. The current phasor at this location produced rated torque, as can be seem in Figure 6-1. It is prudent to ensure that the OC back-EMF is indeed along the positive q-axis by checking its relation to other current angles. Figure 6-3 shows the OC back-EMF and current wave forms for a current angle of $\gamma = 60^o$.

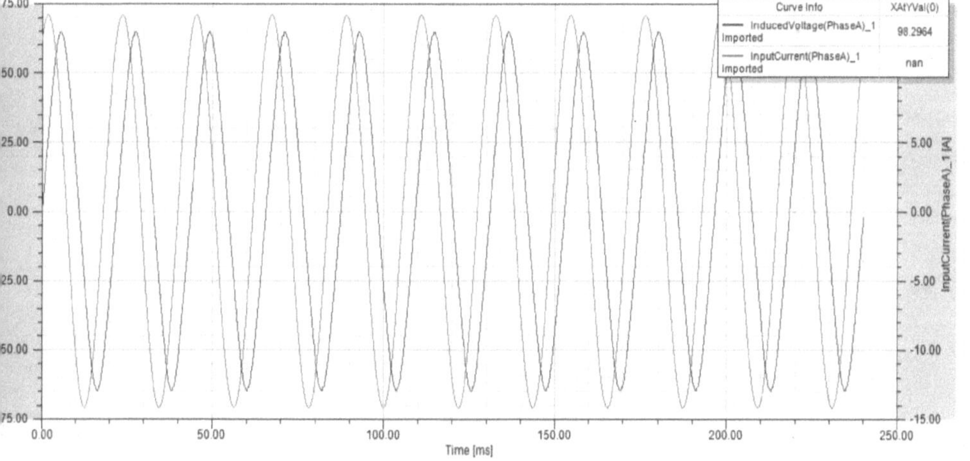

Figure 6-3: Current and b-EMF for γ = 60°

Evidently, the current waveform leads the OC back-EMF by 60°. This indicates that the current phasor is in the 2nd quadrant; or supplying positive q-axis current and negative d-axis current. This is graphically depicted in Figure 6-4.

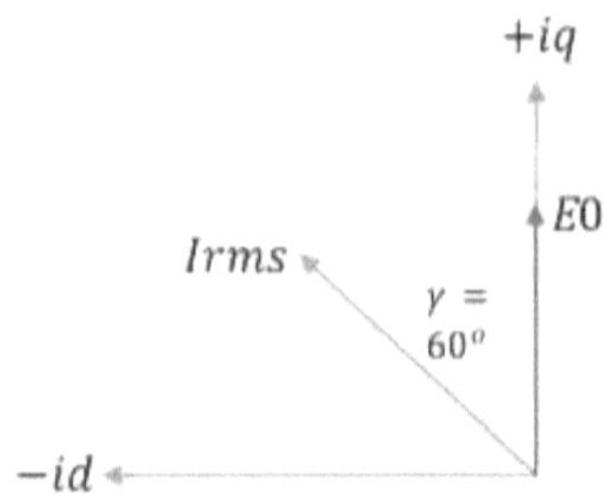

Figure 6-4: Current phasor for $\gamma = 60°$

Under loaded current excitation conditions Maxwell generates an induced voltage waveform. For the sake of alleviating confusion, this waveform should be differentiated between the terminal voltage and the OC back-EMF. Figure 6-5 shows the current excitation induced voltage and the current waveform for a current angle of $\gamma = 0^o$. Close inspection shows that the induced voltage waveform leads the current, albeit by a small phase angle.

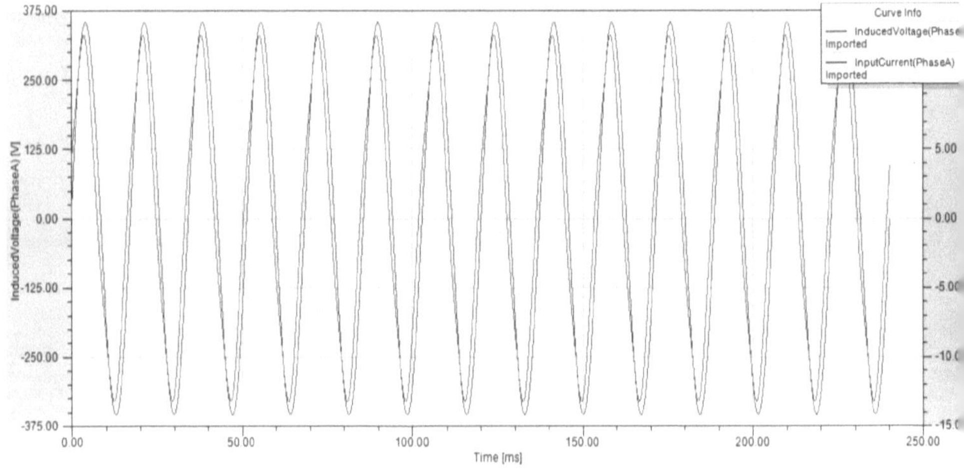

Figure 6-5: Current and induced voltage for $\gamma = 0°$

The $\gamma = 60^o$ case, in Figure 6-6, shows that the induced voltage waveform leads the current, this time by a much larger phase angle.

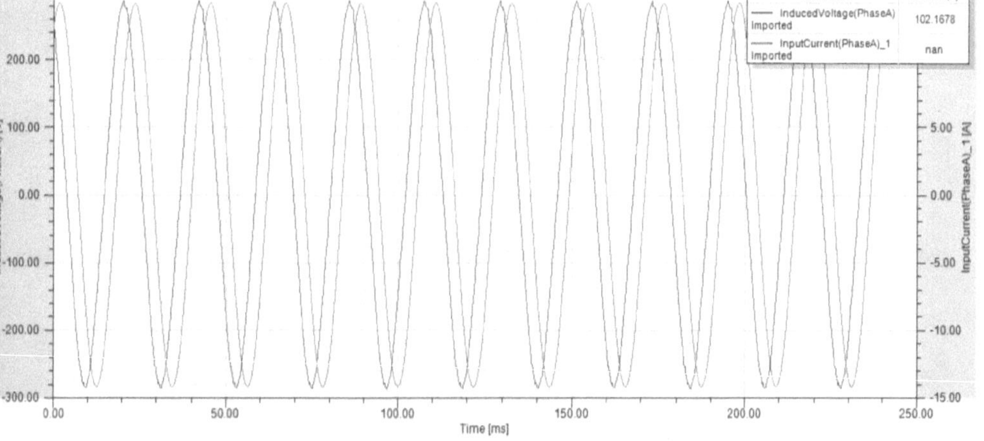

Figure 6-6: Current and induced voltage $\gamma = 60°$

Thus, it can be deduced that the current excitation induced voltage is not the same as the OC back-EMF but can rather be considered to be a representation of the terminal voltage under current excitation.

6.3 Voltage Excitation – Torque Angle Study

The second method is voltage excitation, in which 3-phase purely sinusoidal voltage signals are applied to the machine windings. The voltage signal is given by:

$$v_{exc}(t) = \sqrt{2}V_{rms}\sin\left(2\pi ft + \delta + \theta_{phase}\right) \tag{6.3.1}$$

The machine rated RMS voltage is given by V_{rms} and δ is the torque angle. The torque angle is defined as the angle between the terminal voltage phasor and the OC back-EMF. When a motor is loaded, the rotor flux density field (due to the permanent magnets) lags the stator flux density field (due to the current MMF in the windings). As a result, the terminal voltage leads the OC back-EMF by a phase angle - the torque angle. In generator mode the torque angle is referred to as the load angle.

From the current angle study, it was clear that the OC back-EMF waveform can be used as a reference for the q-axis. Thus, for the subsequent torque angle study, the torque angle will be varied from 0° to 90°. Each angle will yield a different current and voltage phasor in the DQ plane. These current and voltage waveforms will be compared so as to determine the PF at a specific torque angle.

Firstly, it was necessary to ensure that the terminal voltage was indeed leading the OC back-EMF by the torque angle. The following graphs show plots of terminal voltage and OC back-EMF for 0°, 60°, 90°.

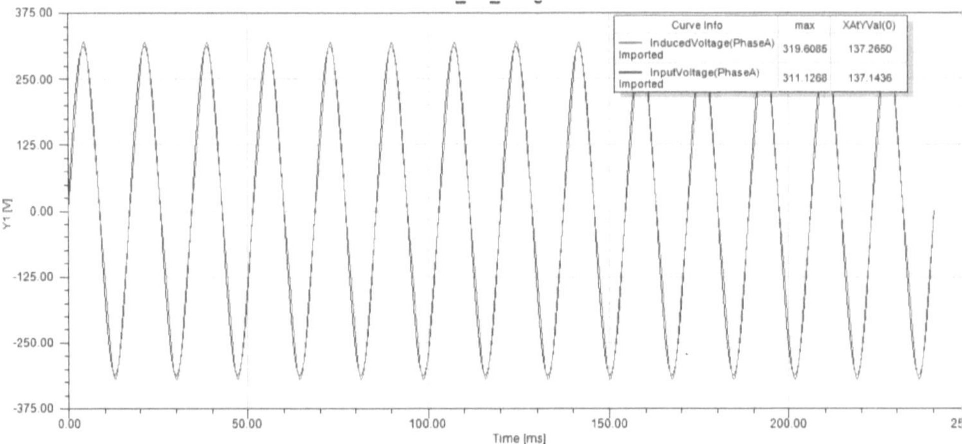

Figure 6-7: Terminal voltage and back-EMF $\delta = 0°$

Clearly for the case of $\delta = 0^0$, the terminal voltage is in phase with the OC back-EMF, as depicted above in Figure 6-7. When the two waveforms are in phase, there is no lag between the rotor and the stator fields, and as result the machine produces its lowest torque. For a torque angle of $\delta = 90^0$, the plot is depicted below in Figure 6-8.

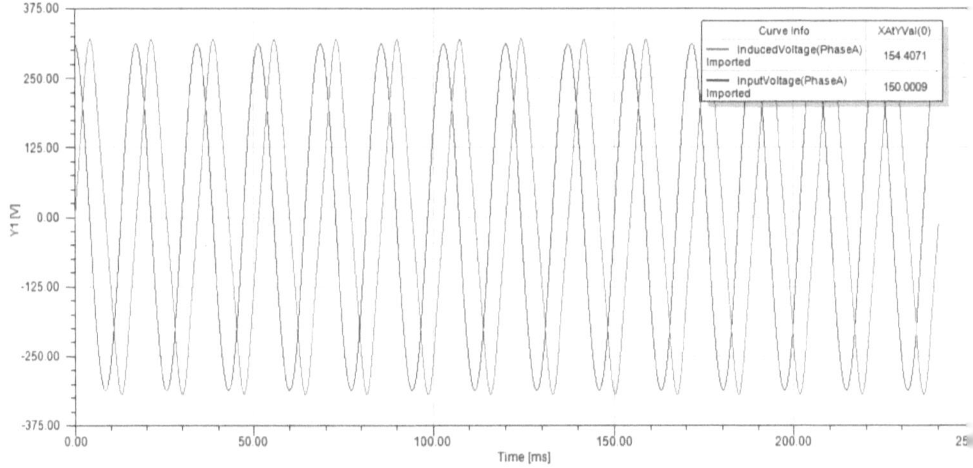

Figure 6-8: Terminal voltage and back-EMF $\delta = 90°$

Evidently the terminal voltage leads the OC back-EMF by 90^0 as expected. This condition yields maximum torque. The variation of torque and current with torque angle is shown in Figure 6-9 and Figure 6-10:

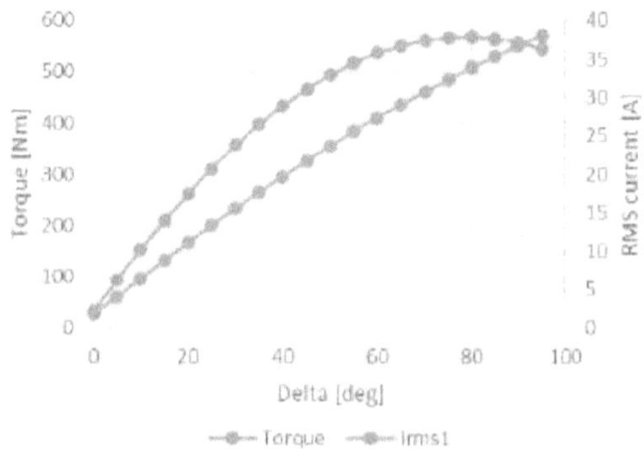

Figure 6-9: FSST torque and current vs torque angle

For the FSST design, as shown above in Figure 6-9, it can be seen that minimum torque is at $\delta = 0^0$ and maximum torque is at $\delta = 90^0$. Rated torque of 230Nm is reached at $\delta = 17^0$ where the current is 10A as designed. The ISSM design torque and current variation with torque angle is shown below in Figure 6-10. It can be seen that minimum torque is at $\delta = 0^0$ and maximum torque is at $\delta = 90^0$. Rated current of 10A is reached at $\delta = 58^0$.

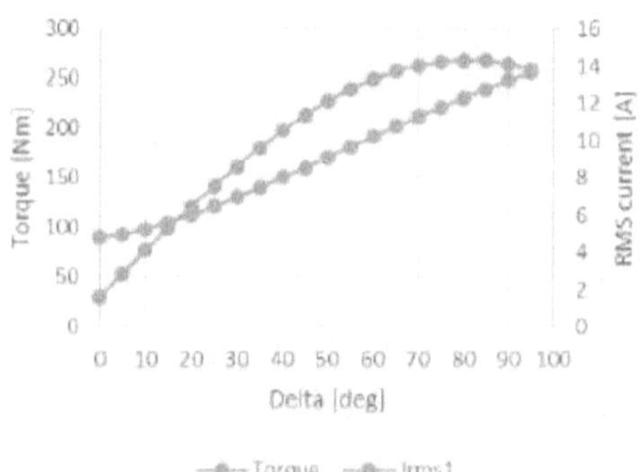

Figure 6-10: ISSM torque current vs torque angle

Notably, the ISSM design yields a higher torque of 245Nm at rated current. This is expected as the ISSM design has a higher pole-ratio than the FSST design, and thus has greater torque density. With regards to the machines' power factor, the terminal voltage waveforms were compared to the corresponding current waveforms at each torque angle. The results are depicted below in figure 11.

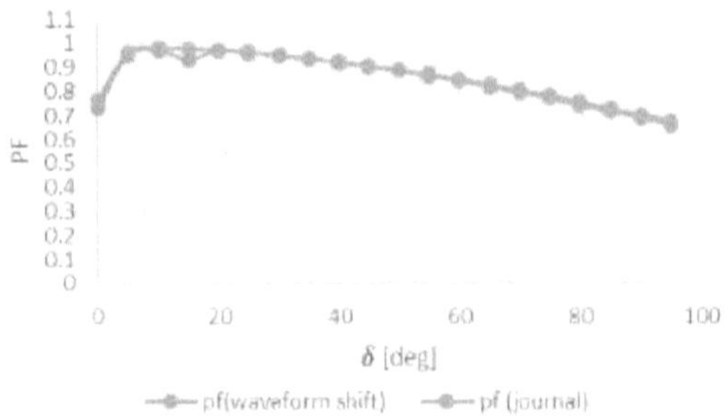

Figure 6-11: FSST power factor vs torque angle

The PF obtained from the angle phase difference between voltage and current waveforms are compared, in Figure 6-11, to the method outlined by [66] and descried in Section 3.6.3. The two methods show very close correlation. As can be seen the PF decreases with increasing torque angle. For the ISSM design, the two methods have the same trend, as seen below in Figure 6-12. However, there is a difference in the magnitude of the PF over the torque angle variation. This difference is due to the lower accuracy of the ISSM simulations. Nevertheless, looking closely at the trend of the PF vs torque angle plot, one will notice that the PF increases to a max around the rated torque angle of $\delta = 58^0$ and thereafter, the PF decreases.

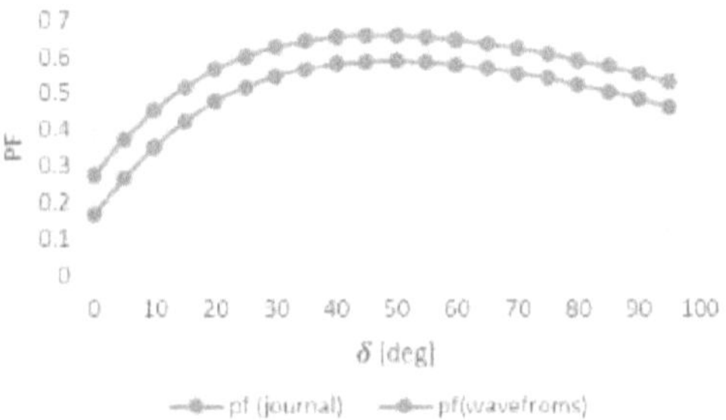

Figure 6-12: ISSM power factor vs torque angle

To summarise, a specific torque angle yields rated torque and current. This is the desired operating point of the machine. The power factor determined from this condition will be the rated power factor. At rated torque angle, the current phasor resides on the q-axis. This is evident in the plot below, Figure 6-13, which shows rated current and OC back-EMF.

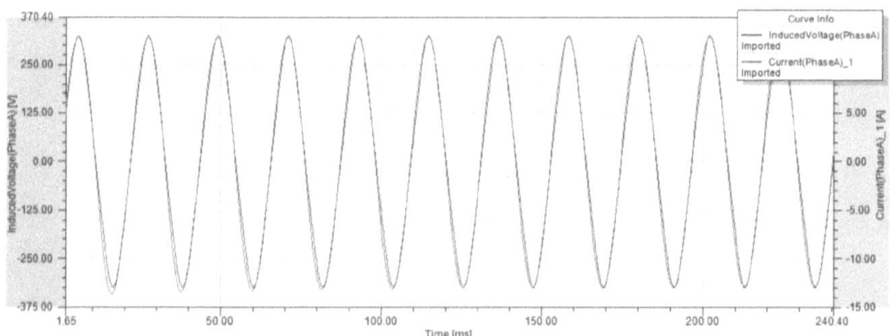

Figure 6-13: ISSM back-EMF & current under voltage excitation at rated torque angle

Evidently, the two waveforms are in phase. Hence the power factors ($\cos(\phi)$) and torque were determined at $i_d = 0$ for both designs. The phasor diagram for $i_d = 0$ operation, neglecting phase resistance is shown below in Figure 6-14. In such circumstances the current phasor, being in phase with the OC back-EMF, is displaced from the terminal voltage by the torque angle such that $\delta = \phi$. From Figure 6-14, X_{FSST} and X_{ISSM} is the synchronous reactance of the FSST and ISSM design respectively. The power factor angles of the FSST and ISSM designs are given by ϕ_{FSST} and ϕ_{ISSM} respectively.

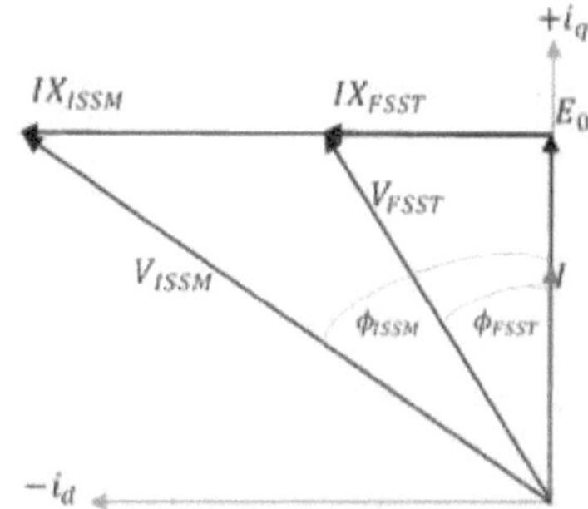

Figure 6-14: Phasor diagram of FSST and ISSM for id operation

Recall that all four designs (ISSM, ISST, FSSM, and FSST) were designed to have the same OC back-EMF (E_0) and rated current (I). The consequences of the Vernier effect are made largely apparent by the phasor diagram. The ISSM design has a higher reactance which is exacerbated by the high pole-ratio and also its larger coil-pitch. The result of this is that the ISSM voltage phasor, V_{ISSM}, is displaced further away from the current phasor resulting in a greater power factor angle, ϕ_{ISSM} , and a greater terminal voltage V_{ISSM}. The ISSM design produces a power factor of $\cos(\phi_{ISSM}) = 0.53$ at a terminal voltage of 390 V_{RMS}. The smaller pole-ratio, smaller coil-pitch and flux-focusing effect of the spoke array in the FSST design further supports the decision to prototype this design. These qualities produce a lower synchronous inductance than the ISSM design. The effect is that the FSST voltage phasor resides closer to the current phasor resulting in a smaller power factor angle, ϕ_{FSST} , and a lower terminal voltage V_{FSST}. The FSST design produces a power factor of $\cos(\phi_{FSST}) = 0.96$ at a terminal voltage of 220 V_{RMS}.

The rated torque waveforms are shown below in Figure 6-15. The plot shows torque waveforms from four different methods, namely; current excitation, voltage excitation, external circuit voltage excitation and a Simplorer co-simulation in generator mode.

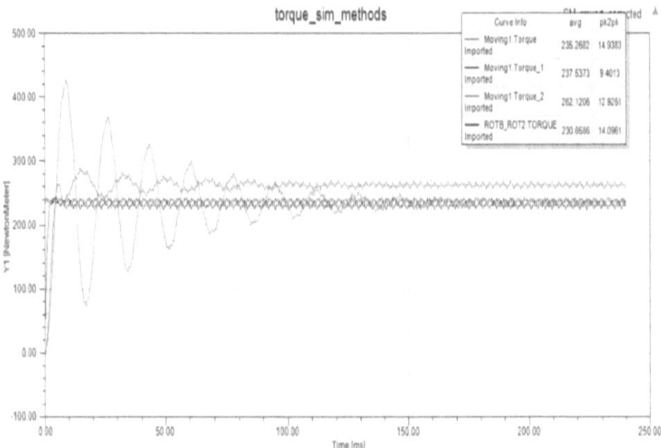

Figure 6-15: FSST rated torque

It is clear the four methods show good correlation. The EC case was simulated with $\delta = 20^0$ (should simulate with for $\delta = 17^0$ for actual report). The voltage and EC methods take longer to reach steady state, as these methods consider the end-winding inductance. Table 6-1 shows the torque characteristics of the four methods. Notably, the current excitation method produces the lowest ripple. This is due to the fact that this method uses a purely sinusoidal input current signal – there are no current harmonics present to exacerbate torque ripple. The other methods all receive harmonics from currents and induced voltages which contribute towards torque ripple.

Table 6-1: FSST torque characteristics

	Average torque [Nm]	Pk-pk torque ripple [Nm]	Torque ripple %	Current THD%	Vind THD%
Voltage	235.2682	14.9383	6.3495	0.78	6.71
Current	237.5373	9.4013	3.9578	0	7.13
EC (20^0)	262.1206	12.9251	4.931	0.69	7.17
Generator	230.6586	14.0961	6.1112	0.49	8.71

The variation of induced voltage THD with torque angle for both designs is shown below in Figure 6-16. As expected, the ISSM design displays lower THD than the FSST design. This is of course due to the higher pole-ratio of the ISSM design.

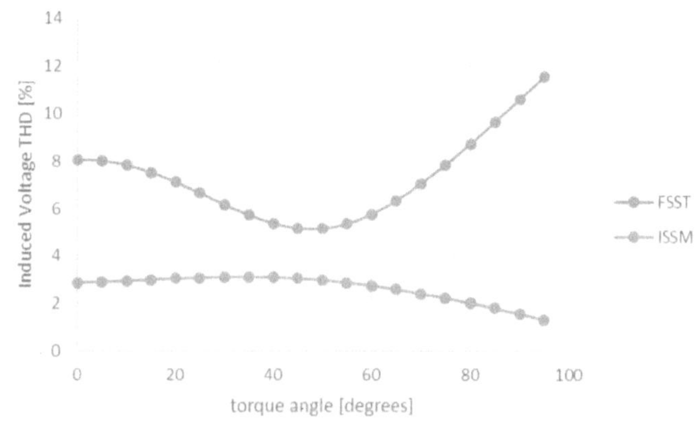

Figure 6-16: Variation of induced voltage THD with torque angle

6.4 Fractional Slot Spoke Type 3D Full Load Simulations

The FSST design was simulated in 3D at the rated torque angle of $\delta = 17^0$ under rated terminal voltage of 220V RMS. The initial 3D model used the winding symmetry (winding periodicity of 2) to simulate only half of the design. The model is depicted in Figure 6-17.

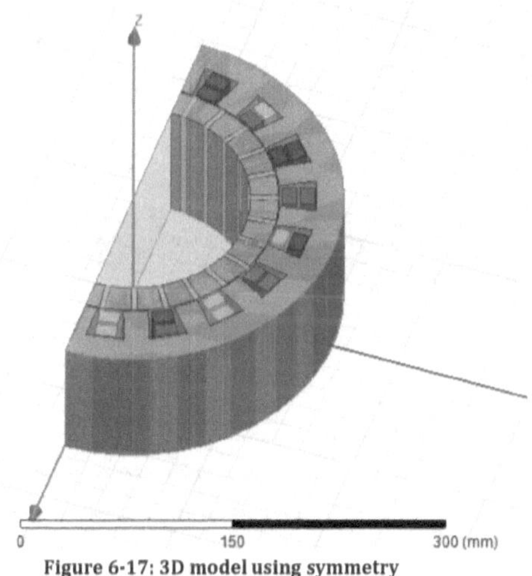

Figure 6-17: 3D model using symmetry

The simulation time was relatively short. A simulation time of 240ms and 255 data samples took 6 hours to complete. However, there was a distinct loss in torque density. The average torque was 216Nm; a 6% loss in torque. The average torque waveform is shown below in Figure 6-18.

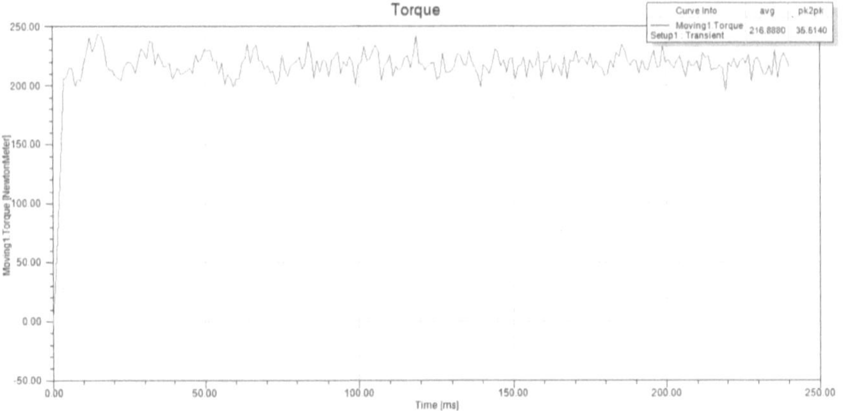

Figure 6-18: Rated torque of the symmetry model

Notably as well, the torque ripple is significantly greater than in the 2D simulations (5-6%). The symmetry model produced a torque ripple of 16.4%. Of course, this type of performance is unacceptable for prototyping. But on the basis of the favourable 2D simulations, it was suspected that the poor performance of the symmetry model was due to the distortion of the flux density at the symmetry planes.

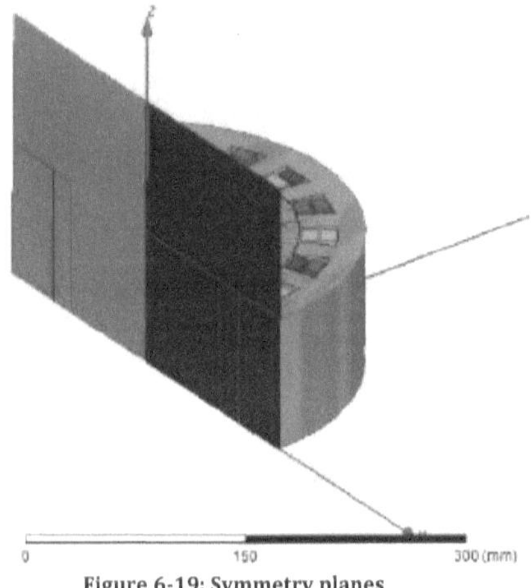

Figure 6-19: Symmetry planes

Thus, a model which uses the full diameter (but still uses axial symmetry) was simulated. The model is depicted below in Figure 6-20.

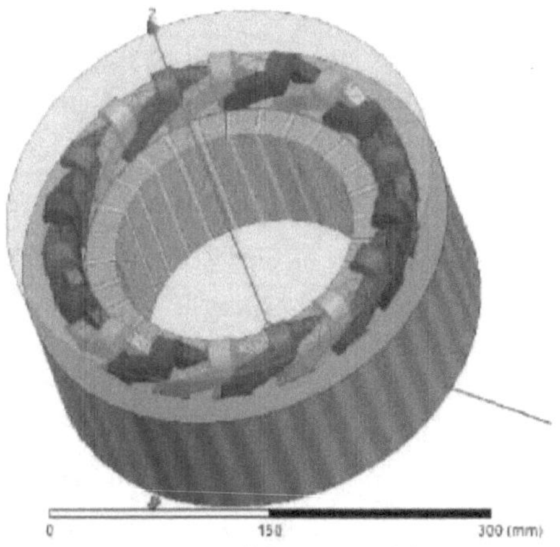

Figure 6-20: Full diameter model

The full diameter model torque waveform is shown below in Figure 6-21.

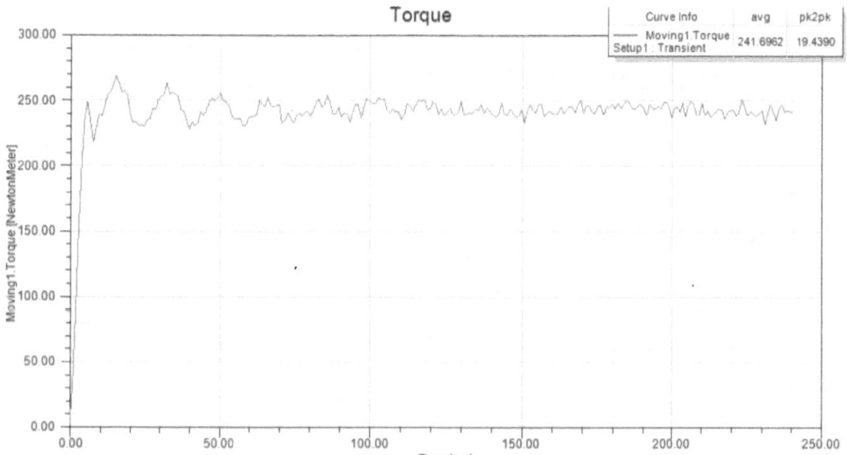

Figure 6-21: FSST 3D simulation rated torque

Evidently, the torque density is higher than in the 2D simulations. This can be attributed to the fact that fewer data samples were taken in the 3D simulations to cut simulation time. The 3D simulation took 3 days to simulate 240ms at 250 data samples. Whereas the 2D simulation took 60min to simulate 240ms at 1000 data samples.

Table 6-2: FSST 2D & 3D torque characteristics

	Average torque [Nm]	Pk-pk torque ripple [Nm]	Torque ripple %	Current THD%	Vind THD%
Voltage	235.2682	14.9383	6.35	0.78	6.71
Current	237.5373	9.4013	3.96	0	7.13
EC (20^0)	262.1206	12.9251	4.93	0.69	7.17
3D EC	241.6962	19.439	8.04	0.47	7.91

This chapter analysed the FSST and ISSM designs in motor mode. This chapter is further evidence that the two topologies will behave as literature dictates under different loads. Explicitly speaking, the power factor is dependent on the nature of the load (the voltage and current angle in this case), and that the FSST case will display a greater power factor than the ISSM case under the same load. Based on this and the outcome of the comparative study, the FSST is chosen prototyping.

7. Fractional Slot Spoke Type Rotor Design

7.1 Original Design Limitations

Considering the active material only (laminations and permanent magnets) as it stands, structurally the FSSM topology is ideal for flux concentration and flux leakage mitigation. The relatively high power factor is a consequence of these features. However, it falls short on its practical realisation capability. The rotor consists of 28 stacks of 399 (199.5mm = 399 x 0.5mm) laminations separated by 28 4mm thick permanent magnets. The complexity lies in attaching these 11200 components to a rotor hub and shaft. Evidently, assembly of such a rotor is a nightmare.

In theory, these flux efficient features can be preserved by using an epoxy adhesive to bond mating components. However, bonds such as these require a non-zero bonding area between bonding faces. Recommended bond-line thicknesses are in the range of 0.0254mm - 0.1778mm. This will consequently add reluctance to the magnetic circuit and reduce the air-gap. Furthermore, there will be 84 surfaces which require adhesive as shown in Figure 7-1. Regulating the bond thickness to the aforementioned range and application of these many surfaces is cumbersome and near impossible. Moreover, uniform bond-line thickness is impossible to maintain along axial length of lamination stack of laminations with varying tolerances.

Figure 7-1: Rotor lamination bonding regions

7.2 Conceptualising Structural Security Features

The preliminary concept includes dovetail wedges which attach the rotor lamination stacks to a stainless-steel rotor hub, and rotor bridges which keep the Permanent Magnets in place. See Figure 7-2 below.

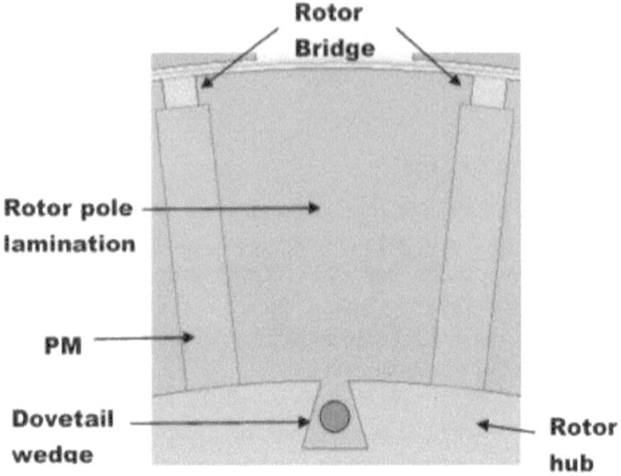

Figure 7-2: Rotor structural security features

Before conceptualising a rotor design, a discussion was had with the manufacturer to determine the design constraints:

- 0.75mm minimum length – due to lack of accuracy in lamination stamping process
- 2mm diameter holes
- M530-50A electrical steel
- Rod thread of 2mm

7.2.1 Dovetail Wedge Concept

It was decided to increase rotor lamination surface area by adding a dovetail wedge feature at the bottom end of the lamination as shown in Figure 7-3. The dovetail wedges will physically keep the rotor laminations in place and eliminate the need for bonding 11000 laminations to a rotor hub. Additionally, this will allow lamination stacks to slide into position on a rotor hub during the assembly. This feature will preserve the required air-gap, as the laminations lie flush on the rotor hub. The dovetail wedges are defined by top and bottom widths and height as shown in Figure 7-3.

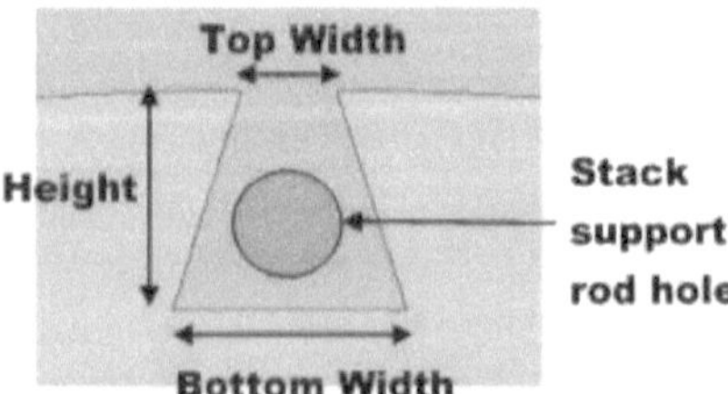

Figure 7-3: Dovetail design parameters

The dovetail wedges will have a stack support hole at the centre for a threaded rod. The rod will keep the lamination stack intact while it is slid into the rotor hub dovetail slots thereby aiding the assembly process. The rod diameter will be determined by the dovetail wedge dimensions and the mechanical integrity of the wedge. See Figure 7-4.

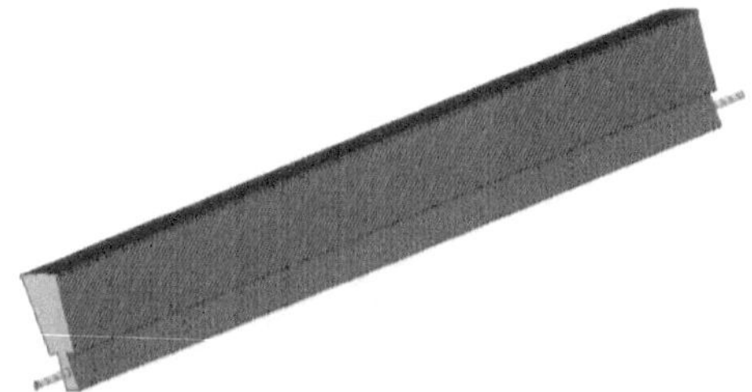

Figure 7-4: Single rotor lamination stack with threaded rod

Note the dovetail slots around the circumference of the rotor hub as depicted in Figure 7-5.

Figure 7-5: Rotor hub depicting presence of dovetail slots

The dovetail wedge is designed to slide into the slots of a stainless-steel rotor hub as shown in Figure 7-6.

Figure 7-6: Insertion of lamination stacks into dovetail slots

Moreover, the dovetails take the shape of a trapezium and thus can be defined in terms of its surface area:

$$A_{dove} = \frac{1}{2}(w_t + w_b)d_h \tag{7.2.1}$$

The significance of the dovetail shape can be seen by analysing the resulting force components acting on the dovetail and consequent reaction forces on the hub.

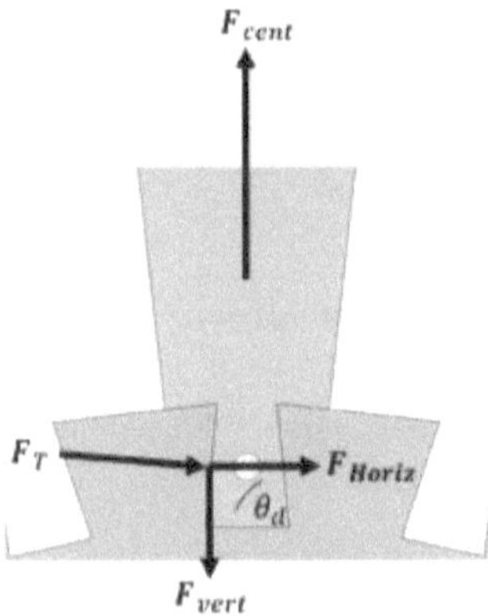

Figure 7-7: Diagram depicting the forces acting on each rotor lamination stack

F_T is the air-gap torque translated to the dovetail wedge exerted by the rotor hub. The shape of the dovetail produces two force components – one in the direction of rotation and one radial force acting towards the shaft. The radial force counteracts the centrifugal force which acts on the rotor lamination stack. The horizontal force acting on the rotor dovetail is given by:

$$F_{horiz} = F_T sin(\theta_d) \tag{7.2.2}$$

The vertical force acting on the rotor dovetail is given by:

$$F_{vert} = F_T cos(\theta_d) \tag{7.2.3}$$

Evidently, this radial force decreases as the angle θ_d increases to 90^0 at which point it is zero. Thus, a nonzero angle creates an inward radial force which is strong enough to mitigate the centrifugal force F_{cent}.

7.2.2 Rotor Bridge Concept

The permanent magnets are held in position by rotor bridges which counter the centrifugal force acting on the magnets. The radial length of the laminations increases to accommodate the dovetail wedge and rotor bridges. The rotor bridges are defined by a height and a width as shown in Figure 7-8.

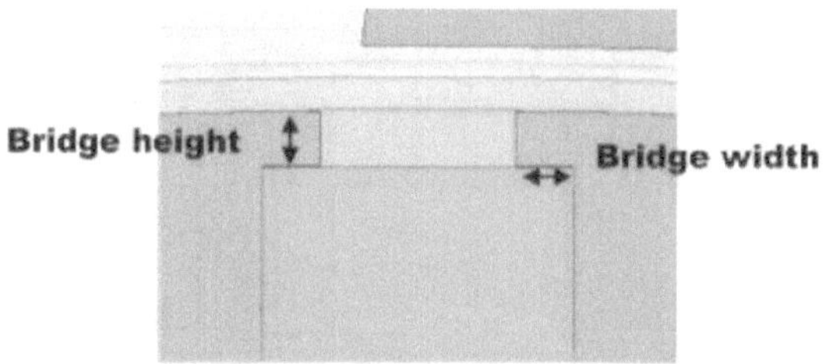

Figure 7-8: Rotor bridge design parameters

7.3 Dovetail and Bridge Analysis

As structural security is the goal, before a magnetic analysis can be done on the aforementioned parameters, mechanical strength constraints need to be imposed on said parameters. Explicitly, an analytical study is conducted on the shear stress (τ_{shear}) on the dovetails and the centrifugal forces (F_{cent}) on the bridges. The forces in question are shown below in Figure 7-9.

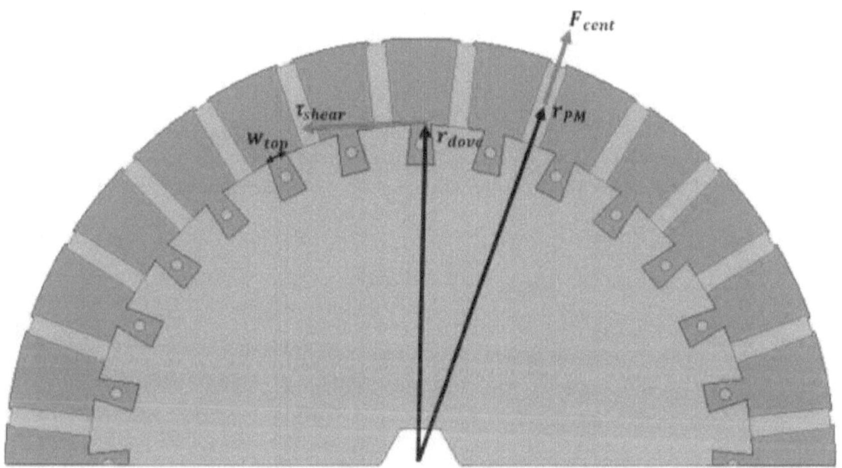

Figure 7-9: Dovetail shearing analysis force diagram

7.3.1 Dovetail Mechanical Constraint

The mechanical integrity of the dovetail wedge is ensured by calculating the maximum shear stress the wedge can sustain before bending occurs. The shear area is shown below in Figure 7-10b, indicated by the orange surface area.

Figure 7-10: Dovetail shearing analysis geometry (a) dovetail top width (b) shearing area

The shear stress acting on the base of the dovetail (indicated by the red line in Figure 7-10a) is given by:

$$\tau_{shear} = \frac{Tr}{J_{pol}} \tag{7.3.1}$$

Where J_{pol} is the polar moment of inertia of the dovetail base –the cross sectional area under shear:

$$J_{pol} = \frac{w_{top} {L_{st}}^3}{3} \tag{7.3.2}$$

Note that τ_{shear} is the allowable shear stress acting on the dovetail before deformation occurs. T is the torque on the dovetail acting to shear the dovetail base. Therefore, the minimum dovetail base width is given by:

$$w_{top} = \frac{3(Tr/\tau_{shear})}{{L_{st}}^3} \tag{7.3.3}$$

The minimum width before bending occurs is calculated to be 1.05mm.

7.3.2 Rotor Bridge Mechanical Constraint

Similarly, a shearing force acts on the rotor bridges. This shearing force is due to the centrifugal force acting on the magnets. The shear area is shown below in Figure 7-11, indicated by the red line, where the area is sheared by half the PM centrifugal force.

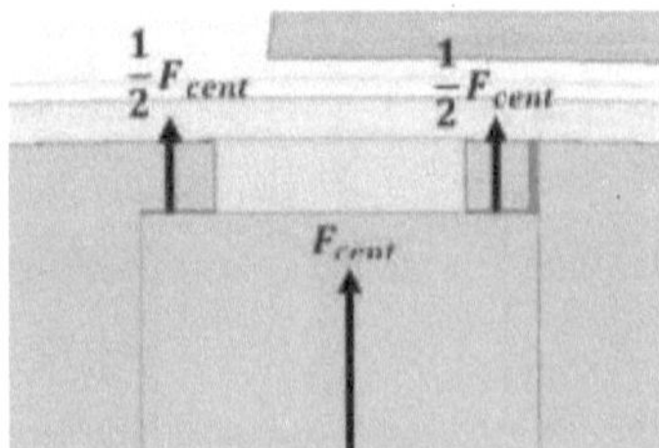

Figure 7-11: Rotor bridge force diagram

The centrifugal force acting on each permanent magnet is given by:

$$F_{cent} = m_{PM}\omega r_{PM} \tag{7.3.4}$$

From the centrifugal force, the shear stress acting on the bridge is calculated:

$$\sigma_{bridge} = \frac{\frac{1}{2}F_{cent}}{A_{bridge}} \tag{7.3.5}$$

Then the minimum bridge height before bending occurs is obtained from:

$$b_h = \frac{F_{cent}}{2\sigma_{bridge}L_{st}} \tag{7.3.6}$$

The minimum width before bending occurs is calculated to be 1.23x10^{-6} mm.

Clearly, the bridge width does not affect the shearing area.

7.3.3 Rotor Bridge Height Magnetic Constraint

Further insight into the bridge height design considerations is obtained from considering the method founded by Bianchi and employed by Li and Lipo et al in [24] in which closed bridges are implemented in the rotor laminations as shown in Figure 7-12:

Figure 7-12: Closed bridge topology employed in some spoke-type designs

In this configuration, the effect of the rotor bridges can be theoretically analysed. The air-gap flux density is given by:

$$B_g = \frac{B_r}{\frac{\pi D_g}{4p_r W_m} + \frac{2\mu_r l_g}{l_m}} \tag{7.3.7}$$

The bridge effect is then taken into account by replacing the remenance B_r by an adjusted value B_{rn}:

$$B_{rn} = B_r - \frac{N_b B_{sat} b_h}{W_m} \tag{7.3.8}$$

N_b is the number of bridges of one magnet; in this case equal to one. Evidently as the bridge height b_h increases, the flux density decreases. This was evident in the open bridge case as well. Clearly, there exists a trade-off between magnet leakage and machining accuracy.

Looking at equation (7.3.8), one might be inclined to mitigate the effect of the bridge height (b_h) on the flux density by increasing the magnet width (W_m). However, the manufacturing must be taken into account. A 4mm thick and 199.49mm long magnet is extremely fragile to fabricate considering how brittle permanent magnets are. Thus, increasing the width of the magnet beyond 20.5 does nothing for improving magnet mechanical strength nor extinguishing manufacturers' complaints.

7.3.4 Bridge Height No-Load Flux Density Analysis

No load flux density plots indicate that the saturation in these regions are unavoidable. The bridge-less case is shown in Figure 7-13.

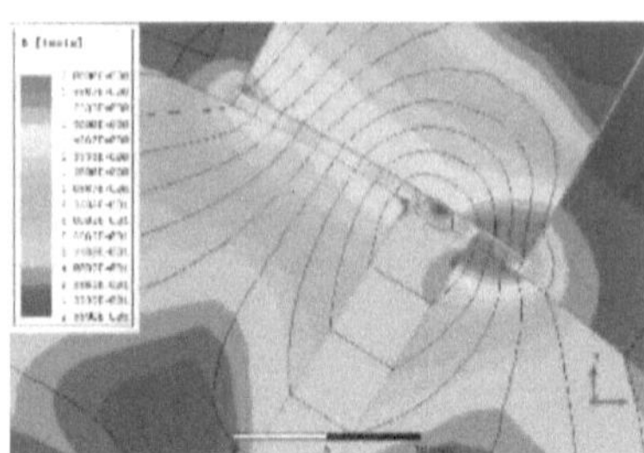

Figure 7-13: Tooth tip saturation when no bridges are present

The tooth tip and bridge saturation was analysed for values between 0.75-1.5mm, the flux density plots of which are shown in Figure 7-14. Evidently, increasing bridge height does nothing to alleviate the saturation in these regions. Saturation is unavoidable in tooth tips. It follows that bridge saturation is inevitable. Varying bridge height does not alleviate saturation. Naturally, air-gap flux density declines with increasing bridge height.

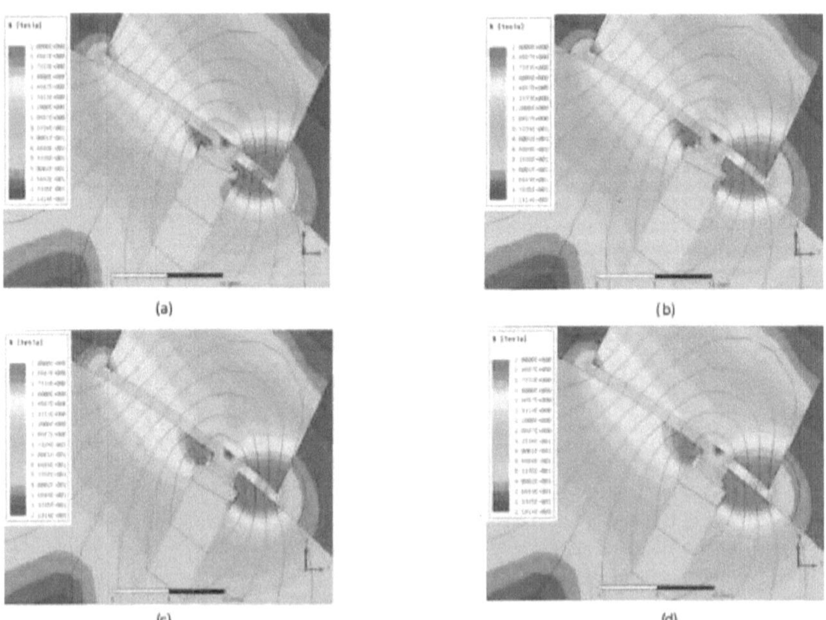

Figure 7-14: Tooth tip saturation for different bridge heights (a) 0.75mm (b) 1mm (c) 1.25mm (d) 1.5mm

Moreover, a parametric study was conducted in which the effect of varying bridge height and bridge width on air-gap flux density was analysed. Finite increments of bridge width are selected according to what is practically realisable; 0.75mm to 2mm where 2mm corresponds to the closed bridge topology. The bridge height is varied from 0.65-1.5mm.

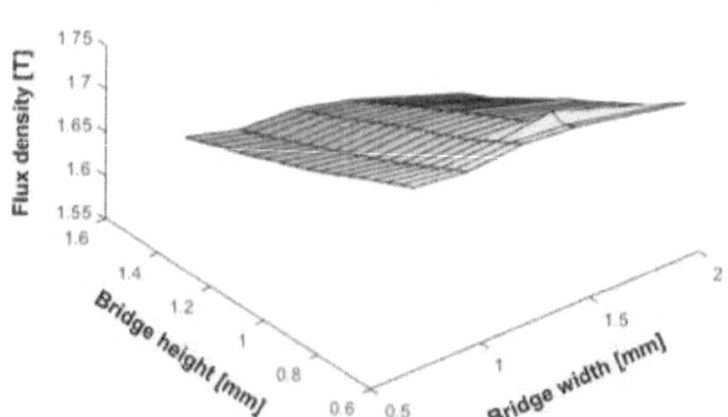

Figure 7-15: Graph of peak air-gap flux density vs bridge height and bridge width

The linear relationship between bridge height and flux density is immediately apparent, especially in the closed bridge case.

Additionally, the Rotor and Vernier harmonic components were analysed using FFT. Considering that the bridges interface with the air-gap, a harmonic analysis was deemed wise to ensure that the presence of the bridges does not eliminate the Vernier effect.

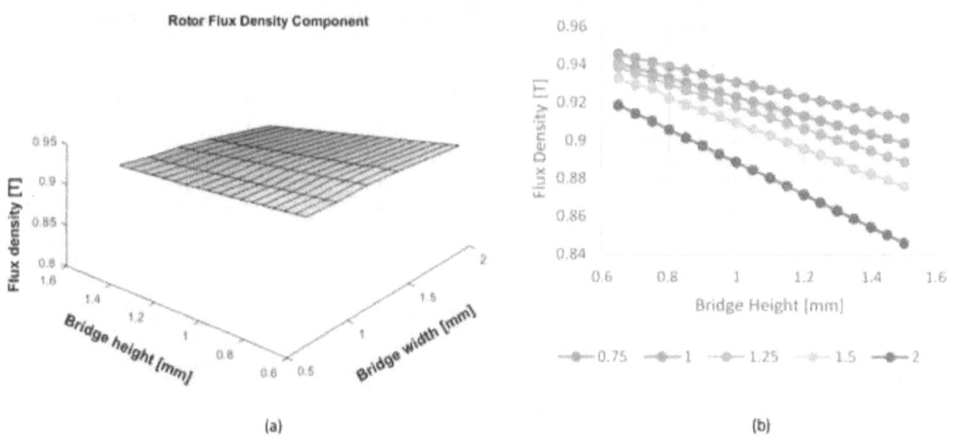

Figure 7-16: Rotor flux density component vs bridge geometry (a) 3D plot (b) 2D plot

Similar to the peak flux density plots, the harmonics diminish linearly with respect to bridge height and width. This is in accordance with equation (7.3.8). The closed bridge case displayed the sharpest decline in flux density. The analysis clearly shows that air-gap flux density diminishes as bridge surface area increases as seen in Figure 7-16b and Figure 7-17b.

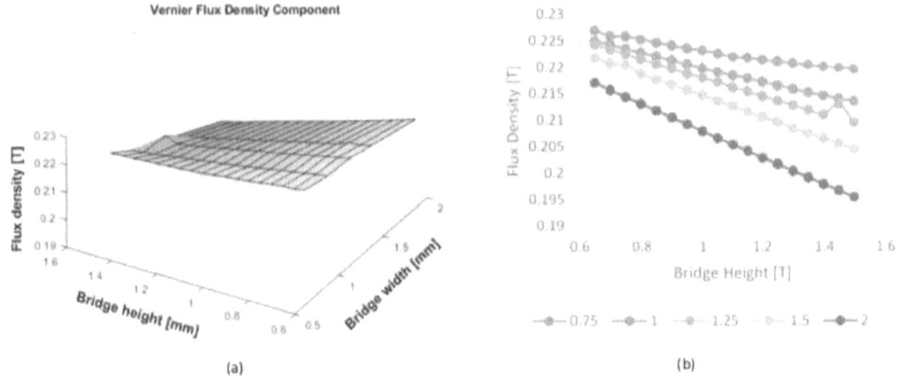

Figure 7-17: Vernier flux density component vs bridge geometry (a) 3D plot (b) 2D plot

These observations can be explained by considering that as bridge width increases, magnet-pole-to-magnet-pole flux leakage increases. As bridge height increases the effective air-gap increases thereby diminishing flux density. The closed bridge (2mm width) topology displayed the poorest flux density performance as well as the steepest decline in flux density with respect to bridge height.

7.3.5 Dovetail No-Load Flux Density Analysis

- Dovetail top width's defining constraint is mechanical strength as the top width carries the machine air-gap torque. As such, the manufacturers placed a 5mm minimum limit on the top width;
- Dovetail bottom width is constrained by pole-pole flux leakage;
- Dovetail height is constrained by pole-pole flux leakage.

A parametric study was conducted in which the effect of varying dovetail top and bottom widths on air-gap peak flux density was analysed. The analysis clearly shows that air-gap flux density diminishes as dovetail surface area increases as seen in Figure 7-18 and Figure 7-19.

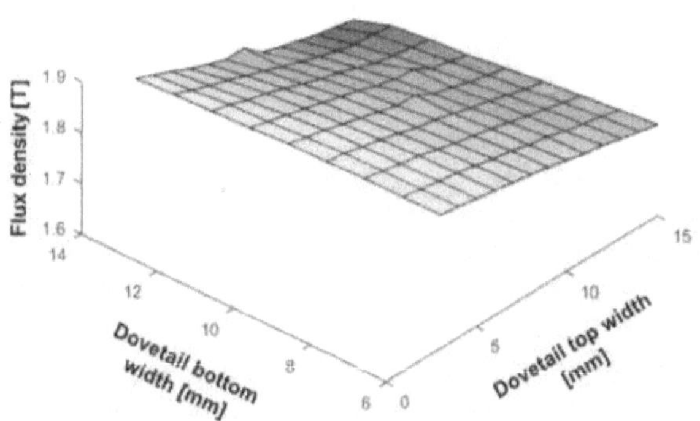

Figure 7-18: 3D plot of the peak air-gap flux density vs dovetail geometry

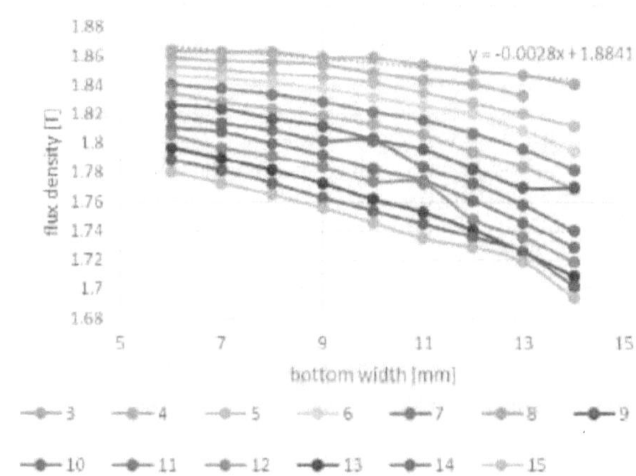

Figure 7-19: 2D plot of the peak air-gap flux density vs dovetail geometry

As dovetail surface area increases, adjacent poles' dovetails reside closer to one another. This creates a shorter path between adjacent rotor poles for flux leakage, as shown in Figure 7-20 indicated by the red arrow.

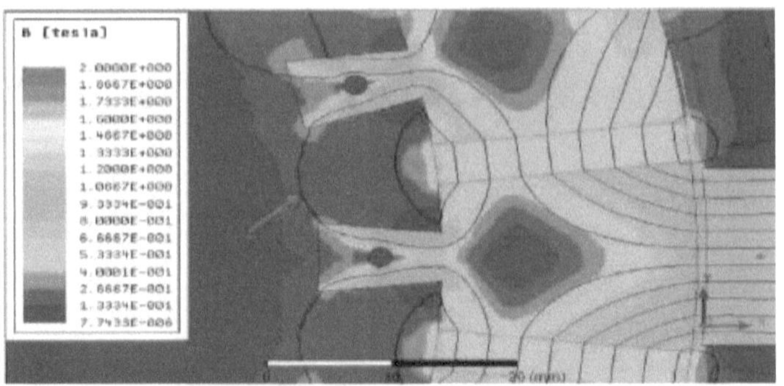

Figure 7-20: Flux density plot illustrating flux leakage between adjacent dovetails

A further constraint was applied to the dovetail top width by limiting it to 5mm. This was suggested by the manufacturer. Considering that it satisfied the shearing analysis in equation (7.3.3), the dovetail top width was kept constant at 5mm in the subsequent flux density analysis.

Further detail is revealed from the Figure 7-19 above by noting the approximate gradients of the curves. The flux density with respect to the dovetail bottom width diminishes at an average rate of 0.0081 [T/mm]. Whereas, the flux density components with respect to bridge height diminish at a rate of 0.039 [T/mm].

A very clear rule-of-thumb is thus evident:

- The smaller the bridges, the less magnet pole-pole leakage
- The smaller the dovetails, the less rotor pole-pole leakage

Evidently, this renders an optimization analysis unnecessary. It is quite clear that an optimization of the aforementioned parameters for power density would yield a solution in which the bridge and dovetail dimensions are zero. This is of course practically unrealisable. Therefore, a more prudent approach was adopted in which a parametric study is conducted to show the decline in air-gap flux density with respect to dovetail and bridge surface area. The dovetail and bridge dimensions will mainly be selected based upon structural integrity and what can be manufactured.

7.3.6 Finalised Dovetail and Bridge Dimensions

It was noted that the decrease in power density was unavoidable. To remedy this, the PM thickness was increased rather than the width or length. This was done to satisfy the manufacturers thereby minimising breakages of brittle PM material during manufacturing or shipping. The PM thickness was increased to 5mm. Final dimensions are shown in Table 7-1:

Table 7-1: Finalised rotor lamination dimensions

Bridge height	0.75mm
Bridge width	0.75mm
Dovetail top width	5mm
Dovetail bottom width	7mm
Dovetail height	10mm
PM thickness	5mm

What follows next is a validation of the above design points. The flux density is compared to the original to confirm that power density reduction is minimised and to ensure the Vernier component is preserved. Thereafter, the torque characteristics are compared in terms of ripple, cogging and magnitude. The power factor is assessed to ensure the prototype aligns with the FSST high power factor characteristic. This is accomplished through a study of the current and torque angles. Finally, the higher steel volume is noted and consequently the iron losses are compared to the original. Ultimately, it is desired that the prototype design behaves as similarly as possible to the original design; in terms of Vernier characteristics and wind power requirements outlined in Section 3.1.

7.4 Increasing the Air-gap Length

The rotor has a mass of 49kg and therefore poses a safety risk due to its high inertia during operation and shear weight during assembly. In order to mitigate these risks, the air-gap length was increased in order to decrease the radial force between the rotor and the stator. This was believed to facilitate ease of assembly thereby mitigating the risk of damaging components and injury to personnel during assembly.

Indeed, during initial testing of the prototype with an air-gap of 0.8mm, distinct metal-on-metal scraping noises could be heard while operating. Moreover, initial operation showed concerning amounts of visible axial movement and eccentric rotation of the rotor. The initial bearing choice of a tapered roller bearing, the high attractive magnetic forces between rotor and stator, combined with the smaller air-gap was deemed to be the cause of this. The prototype was dismantled, and the rotor removed to assess the damage. Figure 7-21 shows clear evidence of the rotor scraping the stator.

Figure 7-21: Stator bore damage due to 0.8mm air-gap

It was decided to resolve this by machining the stator bore to increase the air-gap and to modify the bearing system by using a double row angular contact ball bearing. Based on the initial tests, the tolerance of the tapered roller bearing was not sufficient for this application. The double row angular contact ball bearing provided a tighter tolerance as well as protection in the axial and radial directions. Further reinforcement was achieved by adding a housing for the bearing which secures the bearing in the axial direction.

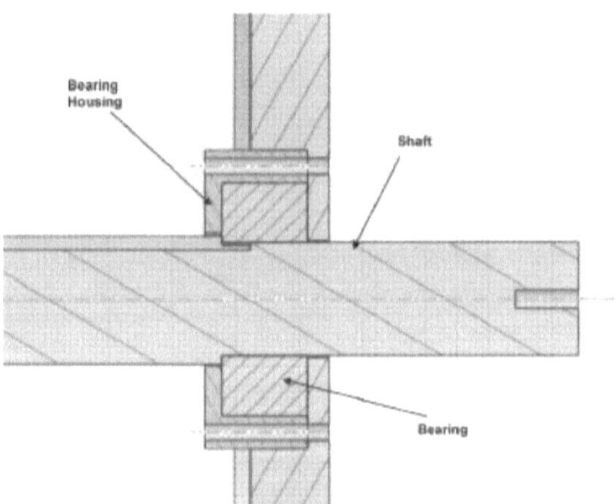

Figure 7-22: Bearing housing modification

7.4.1 Radial Force and the Air-gap Length

This chapter explores the effect of increasing the air-gap length. The air-gap is varied from 0.8mm to 1.7mm in the FEA model and performance parameters are studied. The air-gap variation is limited to 0.9mm to preserve the structural security of the slot wedges which keep the coils securely inside the slots. Due to time and manufacturing constraints, the decision to increase the air-gap was made after the rotor was manufactured and assembled.

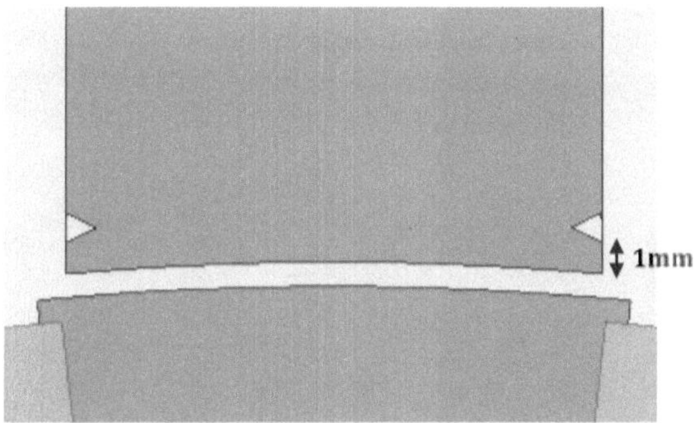

Figure 7-23: Stator tooth notches which hold slot wedges

The radial magnetic force is the main cause of vibration and irregular rotation in electric machines. It results from the magnetic attraction between the magnets on the rotor and the soft iron stator core teeth. The radial force exists in the air-gap between the rotor and stator. And it is this force which poses the greatest hazard to components and personnel during assembly as the rotor is slid into position inside the stator bore.

Radial force density is described by [85]

$$f_{rad} = \frac{1}{2\mu_0}({B_{rad}}^2 - {B_{tan}}^2) \qquad (7.4.1)$$

By integrating around the air-gap circumference and multiplying by the stack length, the air-gap radial force is determined:

$$F_{rad} = \frac{L_{st}}{2\mu_0}\oint({B_{rad}}^2 - {B_{tan}}^2)dl \qquad (7.4.2)$$

Equation (7.4.2), which describes the relationship between radial force and the flux density components, is the motivating reason to decrease the radial force by decreasing the flux density. The stack length is fixed at this stage of the design. Thus, only the stator inner bore is varied in order to parameterise the air-gap length.

Equation (7.4.2) is programmed into the FEA simulator and its dependency on air-gap length is assessed. The air-gap variation is analysed in terms of the same performance criteria used in Section 5.1.

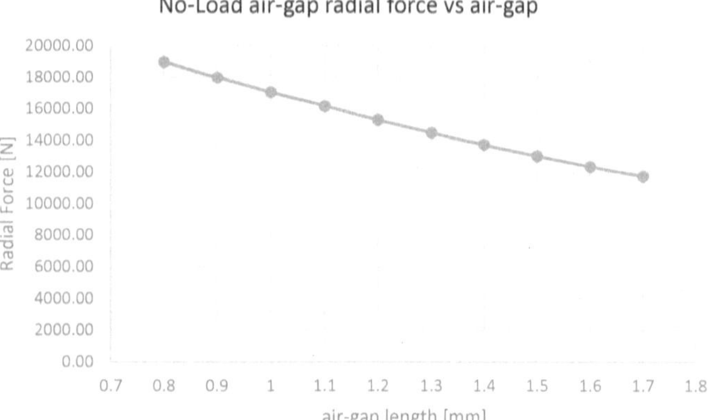

Figure 7-24: Graph depicting the relationship between air-gap radial force and air-gap length in no-load conditions

It is therefore prudent to select an air-gap length which is practical to implement yet would still yield acceptable performance criteria.

Furthermore, the air-gap is chosen to ensure safe operation of the bearing. The safe operation of the bearing is ensured by limiting the radial air-gap such that it is less than the dynamic and static loading of the bearing. The 3207 bearing has a dynamic loading of 40kN and a static loading of 28kN. The air-gap radial force was thus limited to below 50% of the dynamic loading. This ensures safe experimentation and longevity of the machine. As such an air-gap increase of 0.5mm was chosen. Figure 7-24 shows that increasing the air-gap from 0.8mm to 1.3mm decreases the air-gap radial force by 30% to 16.7kN.

7.4.2 Flux Density and the Air-gap Length

The graph in Figure 7-24 shows a clear inverse relationship between radial force and air-gap length. Noting the dependency on flux density in equation (7.4.2), the decline of radial force is expected.

Moreover, consider the equations which constitute the Vernier air-gap flux density which are given in Section 3.1. Equations (3.1.1), (3.1.3), and(3.1.4) indicate an inverse relationship between air-gap flux density and air-gap length. Additionally, it is intuitive to accept that by increasing the air-gap length, the magnetic circuit reluctance will increase, and thus more magnetic energy (for example, larger magnets) will be required to achieve the same power density. To illustrate this the effect of air-gap length on the Vernier and conventional flux density components was studied and shown in Figure 7-25.

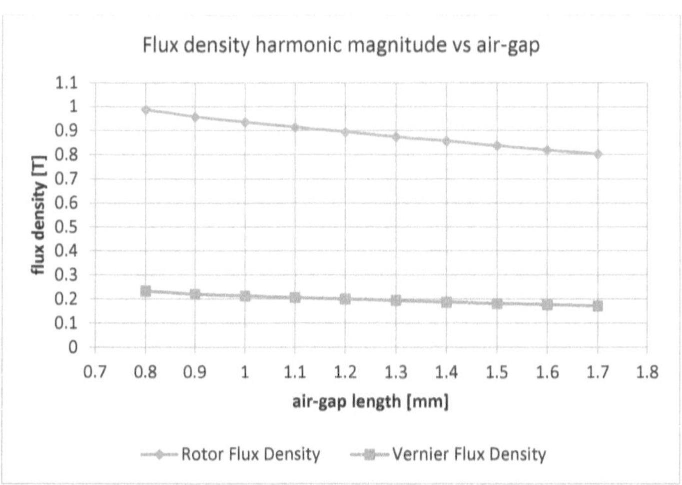

Figure 7-25: Graph of depicting the relationship between flux density harmonics and air-gap length

The graph in Figure 7-25 shows a clear inverse relationship between flux density and air-gap length. The proportionality of the Vernier magnetic loading to back-EMF was described in Section 3.3. It follows that there exists a trade-off with increasing the air-gap length. Decreased power density is unavoidable. However, a benefit of increasing the air-gap is the reduced tooth tip saturation as shown in Figure 7-26 and Figure 7-27.

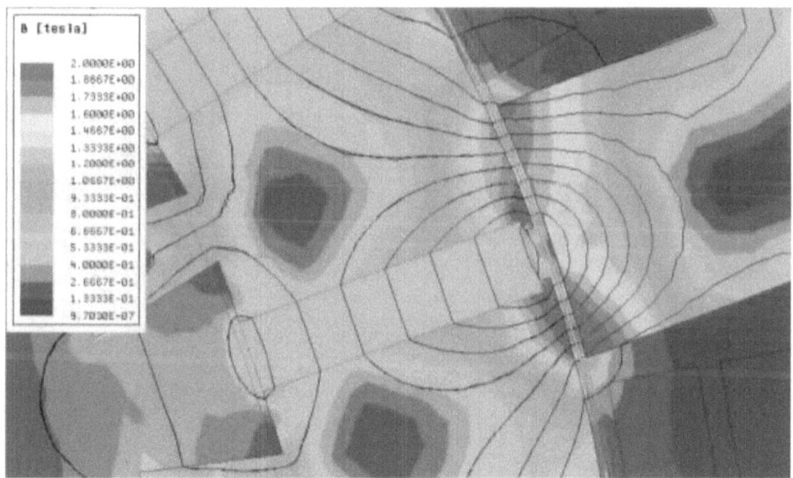

Figure 7-26: Flux density plot of the 0.8mm air-gap design

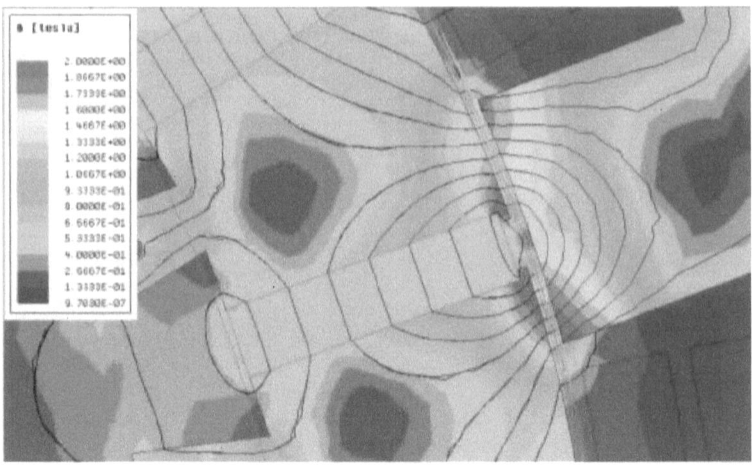

Figure 7-27: Flux density plot of the 1.3mm air-gap design

7.4.3 Full Load Performance and the Air-gap Length

Henceforth, because we cannot have one without the other, it is necessary to assess the extent to which the performance parameters are being sacrificed by increasing the air-gap. To this end, the following analysis studies the effect of increasing air-gap length on torque and power factor and radial force at full-load conditions. The results are shown in Figure 7-28.

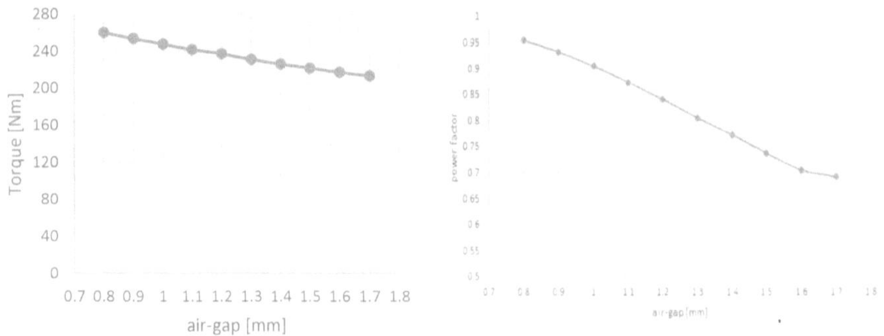

Figure 7-28: Graphs depicting the relationship between torque (left) and power factor (right) air-gap length

As expected, the torque and power factor decline with increasing air-gap length. This is attributed to the lower Vernier magnetic loading as outlined in Section 7.4.2. Evidently, despite the decreased magnetic saturation, the power factor decreased from 0.95 to 0.8 when the air-gap length increased from 0.8mm to 1.3mm. Therefore, it is safe to deduce that the lower leakage flux which Figure 7-27 alludes to is outweighed by the depreciated Vernier magnetic loading.

The air-gap choice of 1.3mm is further reinforced studying Figure 7-28 which shows that the resultant torque will be 231Nm and that the power factor will be 0.8. According to the performance criteria outlined in Section 4.1, the torque magnitude criteria is met. The power factor is significantly reduced, but this is accepted as a necessary sacrifice which can only be amended by adding more PM material.

In order to confirm that the selected air-gap length yields the desired performance criteria, the same analysis is used as in Section 6.3. Consequently, the prototype is simulated in motor mode applying voltage excitation and the torque angle is varied from 0^0 to 90^0. The torque characteristic of the four designs is shown in Figure 7-29.

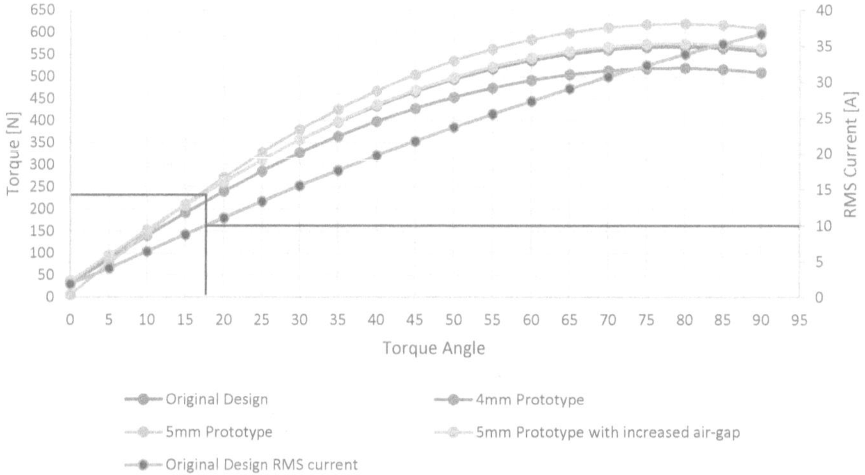

Figure 7-29: Torque characteristic of the four design variations

The figure reveals an interesting tale. By adding the bridge and dovetail features, the torque characteristic drops in comparison to the Original Design. This was expected and analysed in Section 7.3 and put down to the flux leakage in the bridges and dovetails. By increasing the PM thickness to 5mm, the torque characteristic increases to beyond that of the Original Design. The added PM volume provides enough magnetic energy to compensate for the leakage flux caused by the bridges and dovetails. Subsequently, the air-gap increase drops the torque characteristic again, this time matching that of the Original Design.

This is further illustrated in the flux density harmonic spectrum of the four designs as seen in Figure 7-30.

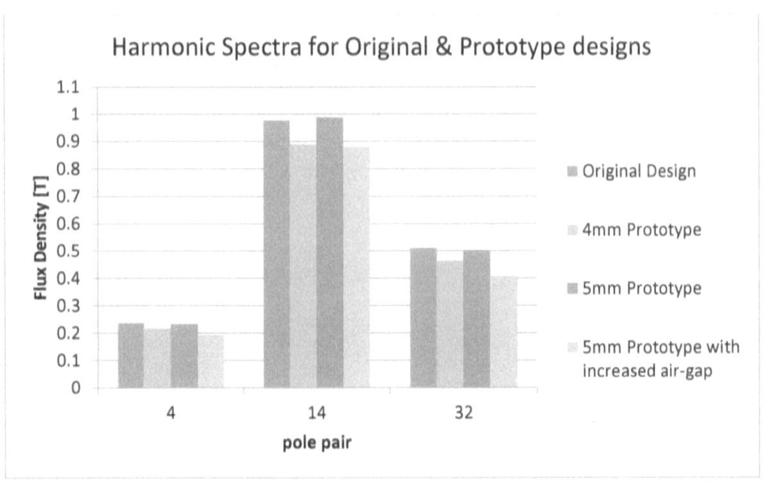

Figure 7-30: Flux density harmonic spectrum showing how the flux density harmonics vary across the design stages

A summary of the performance criteria across the design lifecycle is shown in Table 7-2.

Table 7-2: Table indicating performance parameters of different designs at same operating point

	Steel	RMS Current [A]	torque angle [°]	core loss [W]	Torque [Nm]	torque ripple [%]
Original design	M250-30A	9.338	18	59.8564	231.4424	6.06
Original design	M530-50A	9.37	17	116.7546	231.92	6.46
Prototype design	M250-50A	9.61	17	57.0522	236.2542	6.6
Prototype design	M530-50A	9.64	17	109.569	236.4638	7.1

Evidently, the FSST prototype can deliver similar performance levels compared to the original design in Section 5. The costs involved are lower power factor due to increasing the air-gap length and greater core losses due to increasing the rotor lamination surface area. The costs are deemed small as results indicate the Vernier effect is still present and that the prototype will still deliver desired power output.

7.5 Constructing the FSST Prototype

It was decided to have the prototype manufactured in China. The choice of manufacturer was based on cost, expertise, and quality to ensure compliance with ISO 9001.

7.5.1 Rotor

The complexity of the rotor spoke magnet called for significant mechanical reinforcement of the electromagnetic components. The electromagnetic components being the magnets and laminations, the layout of which can be seen in Section 4.3 , Figure 4-1. The various components which constitute the rotor are shown here in Figure 7-31.

It was critical that all rotor components outside of the magnetic circuit be non-ferrous material. As such, the hub, retainer discs and rods were stainless-steel 304. Soft iron alloys were chosen for the shaft and shaft key for mechanical strength and because these components reside outside of the magnetic circuit.

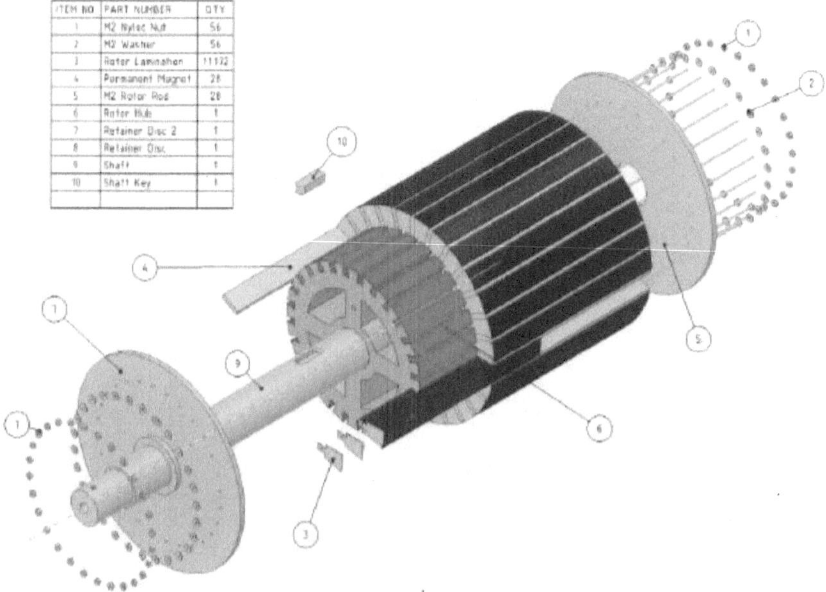

ITEM NO	PART NUMBER	QTY
1	M2 Nyloc Nut	56
2	M2 Washer	56
3	Rotor Lamination	11132
4	Permanent Magnet	28
5	M2 Rotor Rod	28
6	Rotor Hub	1
7	Retainer Disc 2	1
8	Retainer Disc	1
9	Shaft	1
10	Shaft Key	1

Figure 7-31: Exploded view of rotor assembly

The rotor laminations were held together using stainless steel rods, as depicted in Figure 7-4. Stainless steel end plates are attached at either end of the rotor to provide axial reinforcement and compress the 28 rotor lamination stacks.

The shaft key material was selected to ensure that the key could carry the full rated torque of the machine. C45 medium carbon steel was chosen as it offers exceptional tensile strength in the range of 570-700 MPa, which is in fact an over-design.

EN19 (Chinese Standard: 42CrMo) was selected for the shaft material due to its extensive application in rotating shafts. The shaft was heat treated to a surface hardness of 55HRC and internal hardness 28 HRC. This ensured the shaft could support the rotor mass without cold cracking. The shaft is shown in Figure 7-32.

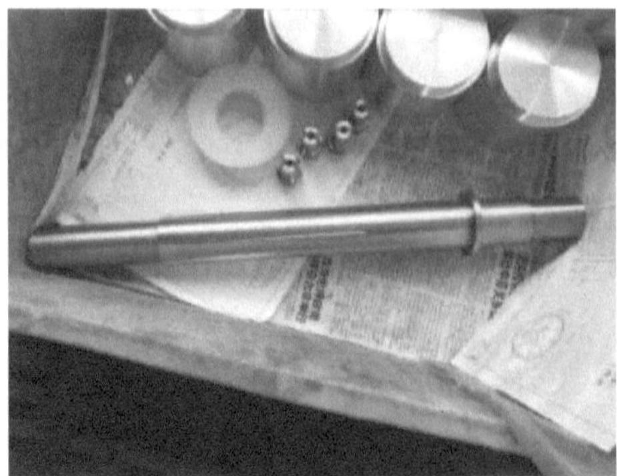

Figure 7-32: Rotor shaft

The manufacturer was tasked with the construction of the rotor. Figure 7-33(left) shows the assembly of the rotor. The Figure depicts the intricacy of the assembly process. The magnets attracted the laminations which therefore required each lamination to be slid into a dovetail slot one-by-one. The rotor finished product is shown in Figure 7-33(right).

Figure 7-33: Rotor - under construction (left) and finished product (right)

After assembly, the rotor was balanced in order to minimise eccentric rotation and to improve bearing longevity. The rotor was balanced to a standard of G6.3. Notches were removed from the rotor, as seen in Figure 7-34, in order to shift the rotor center of mass closer to the axis of rotation.

Figure 7-34: Rotor after dynamic balancing

7.5.2 Stator

With the absence of permanent magnets, the stator construction is relatively less complex than the rotor. The stator assembly is shown in Figure 7-35.

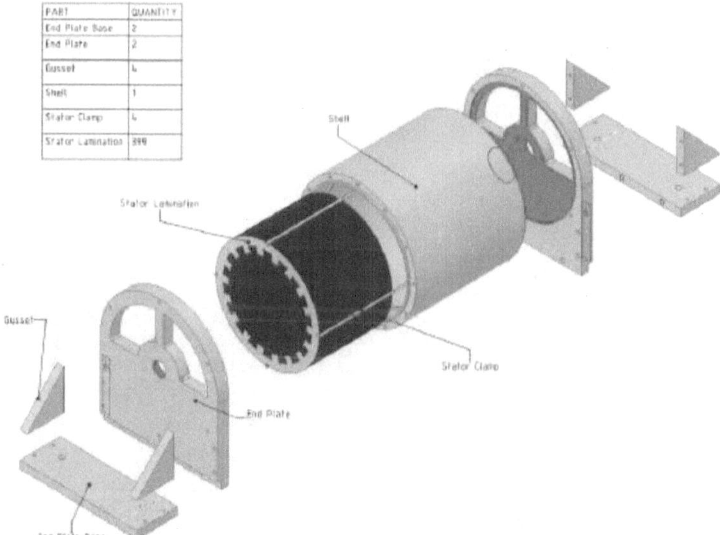

Figure 7-35: Exploded view of stator assembly

The entire stator housing assembly is made from stainless-steel 304 so as not to interfere with the primary magnetic circuit. The stator laminations are M530-50A electrical steel. The stator lamination stack is held together with clamps which run along the stack length. The stator stack is press fitted into the stator shell which houses the entire rotor-stator assembly. The end plates house the bearings. The end plates, base, and gusset form the structure which mounts the entire prototype to the test rig.

Figure 7-36: Stator - under construction (left) and finished product (right)

7.5.3 Experimental Setup

The prototype was setup as shown in Figure 7-37. The prototype was coupled to a 55kW driving induction motor. The induction motor was run in speed control mode. The speed was varied and displayed on the WEG variable speed drive control interface as seen in Figure 7-38c.

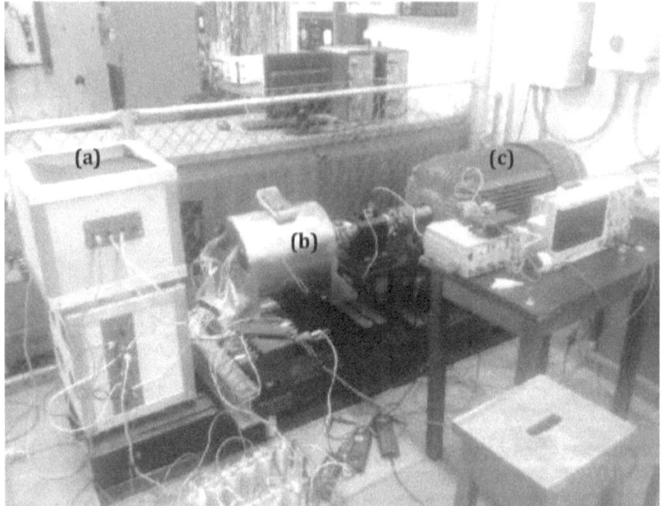

Figure 7-37: Experimental setup – (a) Resistor bank, (b) Vernier PM prototype, (c) Driving induction motor

The in-line torque transducer in Figure 7-38b was used to measure the shaft torque of the prototype. A digital converter converted the transducer signal to analogue milli-volts to be displayed on the Agilent oscilloscope in Figure 7-38a. The transducer has a conversion ratio of 10mV to 1Nm. Throughout testing, the winding temperature was monitored using a thermocouple and a digital thermometer in Figure 7-38d. The temperature was monitored to ensure temperatures do not exceed safe operating limits which would damage winding insulation and demagnetize permanent magnets.

The prototype was loaded using 3-phase resistor banks. Different loading points were obtained by switching combinations of 48Ω resistors in series and parallel. The Agilent oscilloscope recorded current, voltage, and torque waveforms and discrete data points.

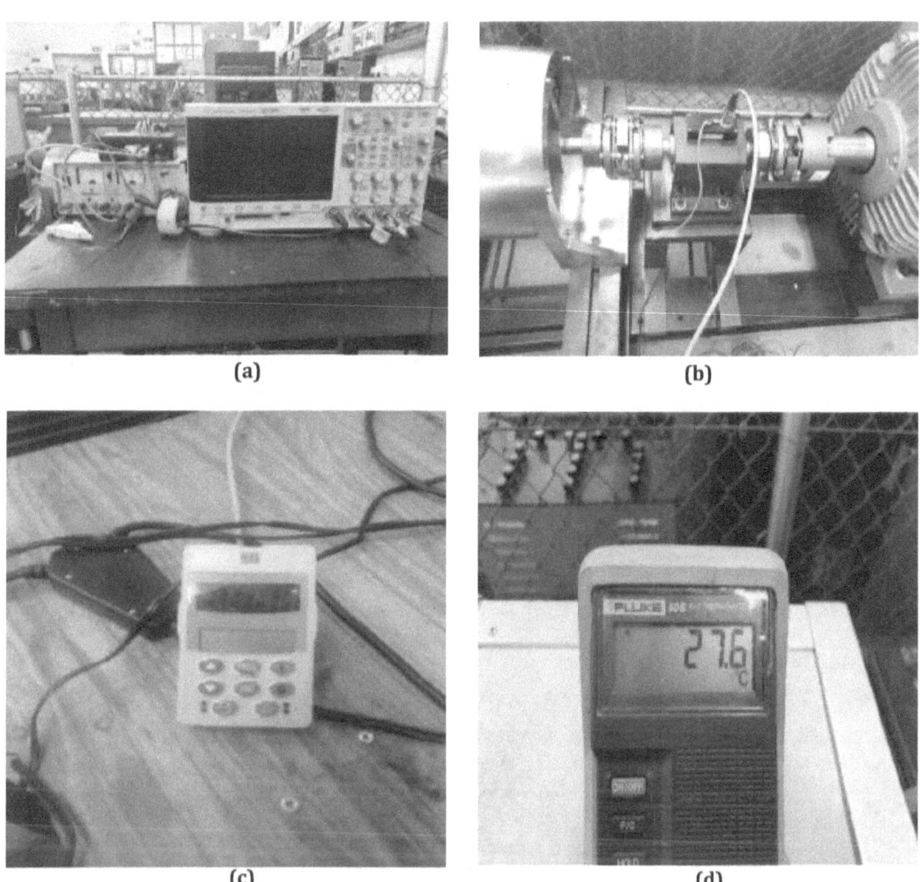

Figure 7-38: Data capturing equipment - (a) Oscilloscope, (b) Torque transducer, (c) Speed control interface, (d) Digital thermometer

8. Experimental Results

This section presents the no-load, cogging torque, and full-load experimental results.

8.1 Winding Resistance and Inductance

The winding resistance and inductance was measured using an LCR meter. The resistance and inductance for one phase belt is shown in Figure 8-1. The average resistance was found to be 0.94 Ω, and the inductance 5.5mH for the two phase belts.

Figure 8-1: Single phase belt resistance and inductance measurement

These values do not agree well with analytical nor FEA results. This was the first indication that machine was not as designed.

8.2 No-load Tests

The no-load tests were conducted by leaving the generator terminals open circuit and connecting directly to the measuring device. An oscilloscope was used to measure the open-circuit back-EMF. The no-load test was used to estimate no-load losses and back-EMF, as well as cogging torque.

8.2.1 Back-EMF

The machine was run at varying percentages of rated speed (250RPM). This enabled the assessment of back-EMF magnitude dependency on speed. Firstly, the phase back-EMF at rated speed is shown below here in Figure 8-2.

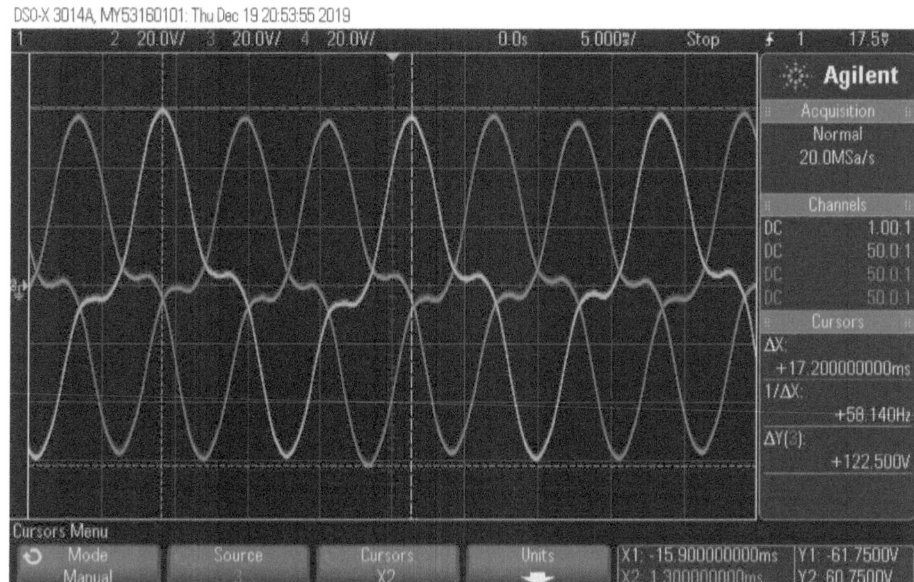

Figure 8-2: Phase back-EMF at 250RPM

It can be seen that the machine is generating balanced voltage in magnitude and phase. However, the waveform is not sinusoidal as expected from the simulations in the preceding paragraphs.

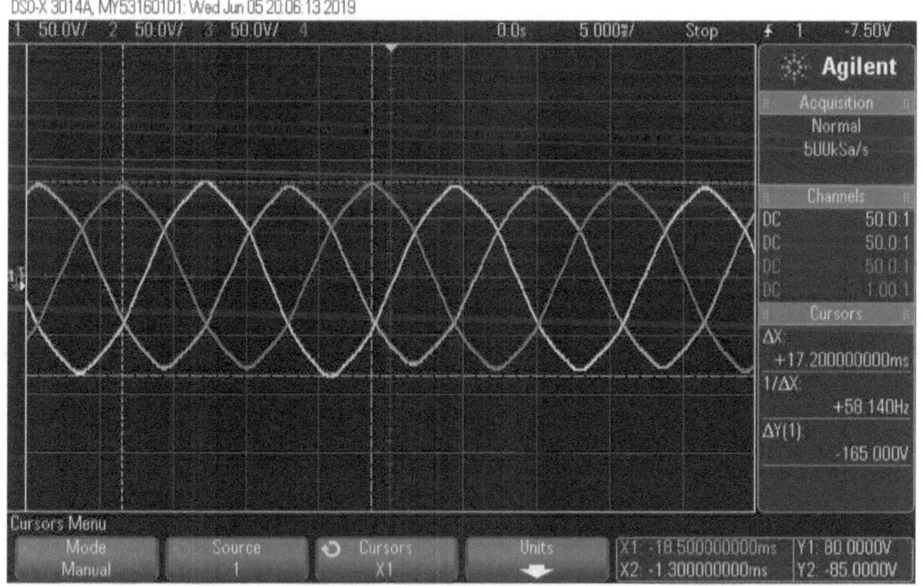

Figure 8-3: Line back-EMF at 250RPM

Furthermore, the magnitude is lower than expected. It is clear that there is a significant harmonic presence in the back-EMF waveforms. What follows is an assessment of the harmonic spectra of the back-EMF. The harmonic spectrum of the back-EMF phase waveform in Figure 8-2 is shown below in Figure 8-4.

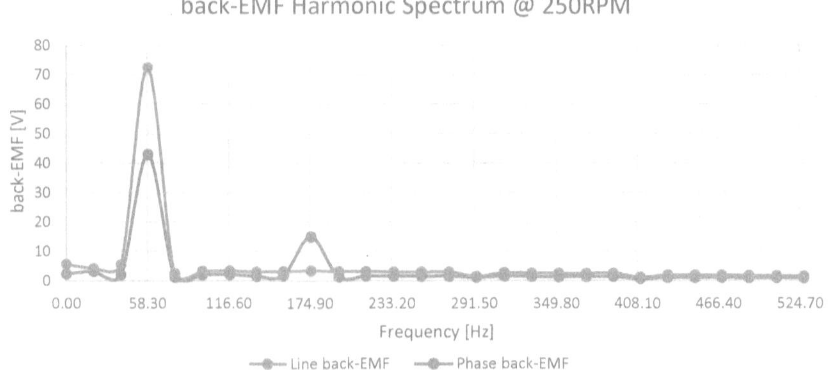

Figure 8-4: Harmonic spectrum of voltage at 250RPM

The harmonic spectrum reveals that the third harmonic is the dominant parasitic harmonic. The fundamental harmonic at 58.3Hz is 43V. The third harmonic at 174.9Hz is 15V. The fifth harmonic at 291.5Hz, seventh harmonic at 408.1Hz, and ninth harmonic at 524.7Hz are so low it is safe to consider them inconsequential. The total harmonic distortion of the phase back-EMF is calculated using equation (8.2.1) which is the voltage variation of equation (3.6.13).

$$THD = \frac{\sqrt{\sum_{n=2}^{\infty} V_n{}^2}}{V_1} \tag{8.2.1}$$

The THD of the line back-EMF is 6% while the THD of the phase back-EMF is 35%. The high THD of the phase back-EMF is due to the high the 3rd harmonic. The 3rd harmonic is not present in the line back-EMF due to the elimination of triplen harmonics in 3-phase line-line connections. The absence of other parasitic harmonics in both line and phase back-EMF waveforms is due to the distribution of the fractional slot windings.

Furthermore, consider the outcome of the paper [16]. The paper shows that the line back-EMF of a Vernier machine is inherently sinusoidal due to the magnetic gearing effect and gearing ratio.

The no-load back-EMF as a function of shaft speed is shown in Figure 8-5. The linear relationship of voltage with respect to shaft speed is evident.

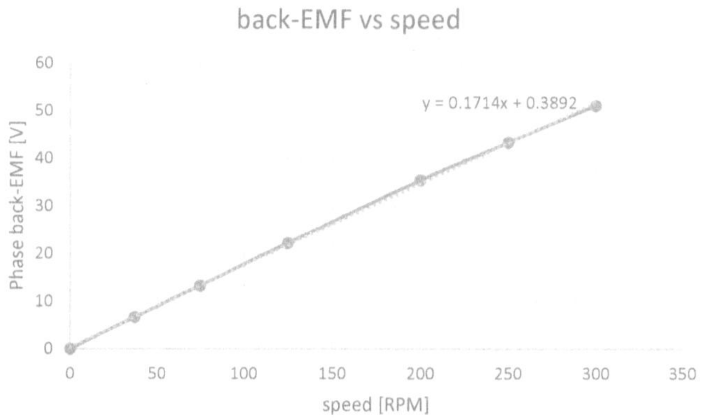

Figure 8-5: back-EMF relationship with shaft speed

The flux linkage of the Vernier prototype can be deduced from Figure 8-5 by noting that voltage is equivalent to the rate of change of flux linkage with respect to time as per equation (3.3.8) in Section 3.3. The prototype flux linkage is thus equal to the gradient of the line in Figure 8-5; 0.1714 Wb-turns.

The prototype was thus re-simulated with the defects in mind. The number of turns per phase was reduced from 180 to 90. The stator and rotor were simulated as solid entities rather than laminated. The comparison of the phase back-EMF harmonic spectra is shown in Figure 8-6.

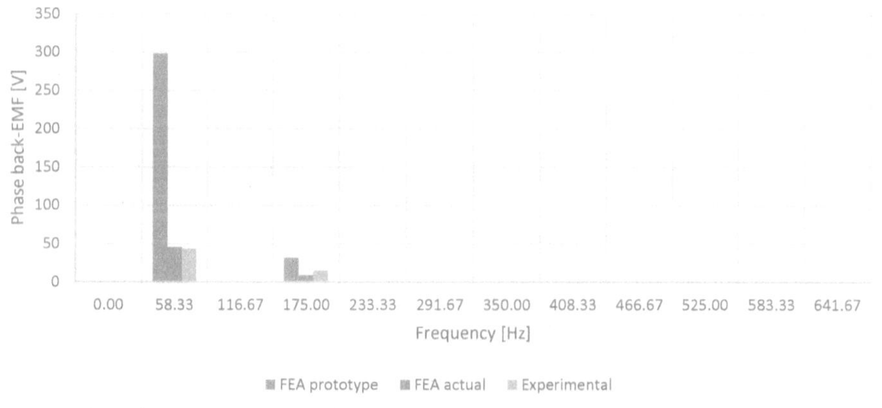

Figure 8-6: Harmonic spectra of FEA and experimental back-EMF

By delaminating the stator and rotor, the increased eddy current loss brings about a larger 3rd harmonic. The 1st harmonic of the FEA matches the experimental back-EMF quite well.

8.2.2 No-Load Losses

No-load losses are the sum of the core losses and the friction and windage losses. The core loss can be determined by subtracting the friction and windage loss from the total no-load losses. Alternatively, the no-load losses can be determined by subtracting the copper losses from the total losses and noting equation (3.11.5) such that:

$$P_{NL} = P_{loss} - P_{Cu} = P_{shaft} - P_{elec} - P_{Cu} \tag{8.2.2}$$

The friction and windage loss is usually measured experimentally by running the machine with a dummy (non-magnetic and non-conductive) stator inserted [86], [87]. Unfortunately, owing to time and manufacturing constraints, a dummy stator was not an option in this analysis.

Friction and windage loss was therefore calculated deterministically using the bearing frictional coefficient k_{fb} and equations (3.10.1) and (3.10.2) respectively in Section 3.10. The frictional coefficient was obtained from the manufacturer datasheet. The friction and windage loss is then subtracted from the measured no-load losses to get the iron loss.

Equations (3.9.1) and (3.9.2) in Section 3.9 indicates that the core losses are proportional to the frequency, and thus the shaft speed, of the machine. The core loss at varying speed range was analysed to assess this relationship. The result is shown in Figure 8-7.

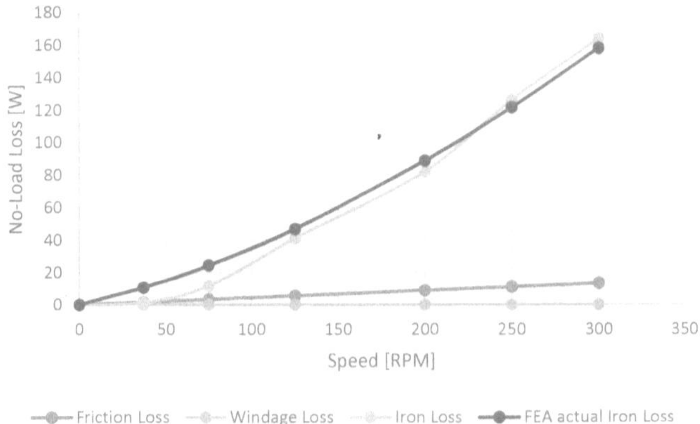

Figure 8-7: No-load losses with respect to shaft speed

This quasi-analytical approach is deemed appropriate here due to the prototype low operating speed and comparatively small rotor size. Figure 8-7 reveals that the FEA model using actual parameters matches the experimental iron loss reasonably well.

8.2.3 Cogging Torque

The cogging torque measurement was done at a much lower speed than rated speed in order to eliminate the effect of no-load losses on the readings. The cogging torque over one rotor rotation at 37RPM is shown in Figure 8-8.

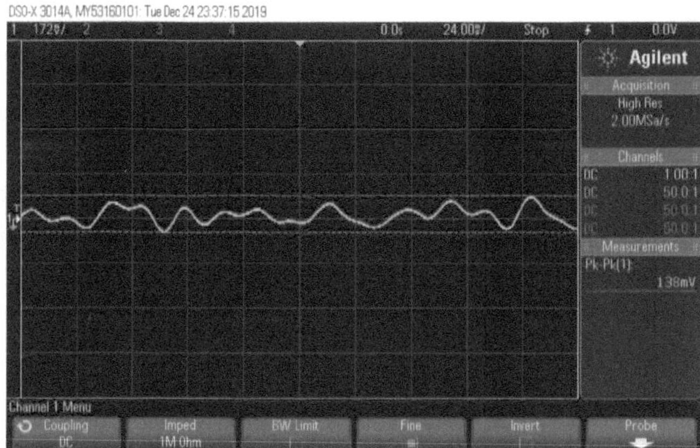

Figure 8-8: Cogging torque at 37RPM

The cogging torque has peak-peak magnitude of 13.8Nm.

Equation (3.7.5) indicates that the cogging torque is proportional to the air-gap flux density. Thus, one may be lead to believe that this is the cause of the high peak-peak cogging torque in figure. Moreover, consider that the back-EMF is also proportional to the air-gap flux density. Yet, the back-EMF in figure is lower than expected. It follows that a large cogging torque and low back-EMF is unforeseen and that the flux density cannot be the cause for the high cogging torque. The issue undoubtedly lies with the windings

The high cogging torque can be attributed to the fact that no skewing or tapered teeth was implemented. The open slot structure which emphasises the variable reluctance of the slotting effect (to facilitate the Vernier effect) thereby maximising cogging torque. And by equation (3.7.5), a high variable reluctance will yield a high cogging torque.

8.3 Load tests

The full-load tests were conducted by connecting 3-phase resistors in series to the machine windings. The resistors were varied from 50Ω to 16 Ω per phase. As such the load, and thus the current, on the machine could be varied. This allowed analyses of various machine operating points. Additionally, capacitor banks were connected in series to the resistors to assess the effect of non-linear loading.

8.3.1 Load Conditions

The prototype was loaded by connecting three 48 Ω resistors in parallel such that the total load was 16 Ω. The machine was then run at rated speed of 250RPM.

The voltage waveform and harmonic spectrum is shown in Figure 8-9. The 3rd harmonic is once again present being the dominant parasitic harmonic. The voltage waveform has a THD of 37%.

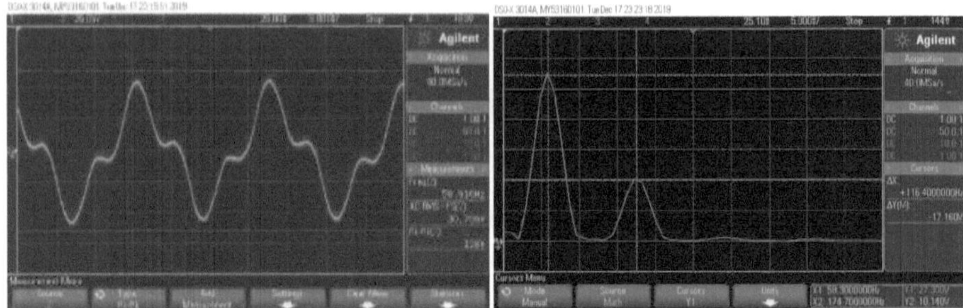

Figure 8-9: Terminal voltage waveform (left) and harmonic spectrum (right)

The current waveform and harmonic spectrum is shown in Figure 8-10. Consequently, the voltage 3rd harmonic yielded a 3rd harmonic in the current waveform. The current waveform has a THD of 28%.

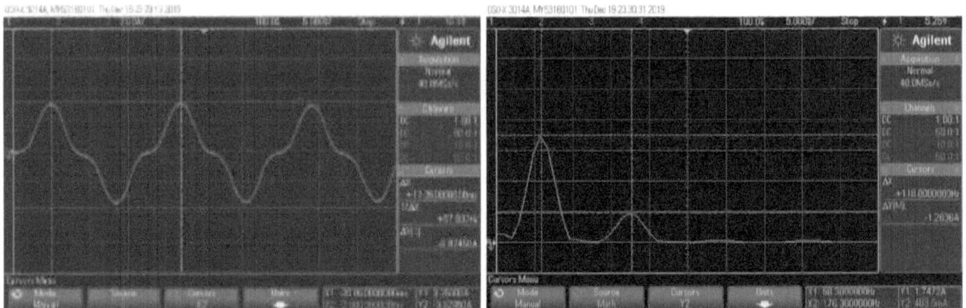

Figure 8-10: Load current waveform (left) and harmonic spectrum (right)

Due to the high THD, it follows that the torque ripple will be relatively large as well. The torque waveform is shown in Figure 8-11. The torque magnitude is 16.4Nm and the percentage torque ripple is 49%.

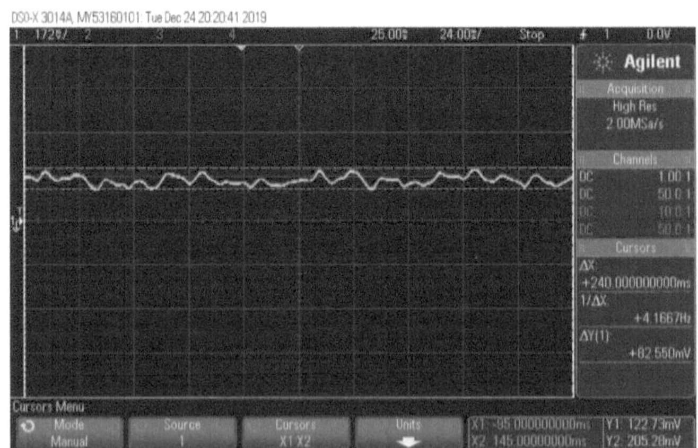

Figure 8-11: Torque waveform

8.3.2 Voltage Regulation

The per phase voltage regulation at different shaft speeds for a purely resistive load is shown in Figure 8-12.

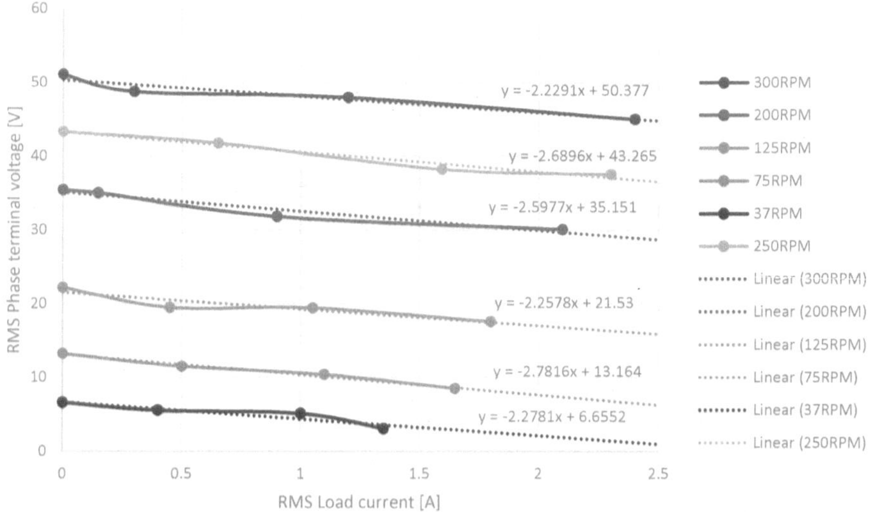

Figure 8-12: Per phase voltage regulation at different shaft speeds

The gradient of the Voltage vs Current curves in figure varies from 2.2V/A to 2.8V/A over the test speed range. The gradient at rated speed is 2.7V/A. The steepness of the graphs, and thus the poor voltage regulation, is due to the large winding resistance and low flux linkage.

8.3.3 Power Factor for a Non-Linear Load

The power factor is not as straight forward to extrapolate from measurements of the prototype in generator mode. The power factor is dependent on the nature of the load. The power factor using a 105nF capacitor in parallel with a 16 Ω resistor is thus determined using the method in Section 3.6.3.

In generator mode, and in particular this prototype, the harmonic content is too severe to use the phase displacement method. Figure 8-13 shows the voltage and current waveform phase relationship.

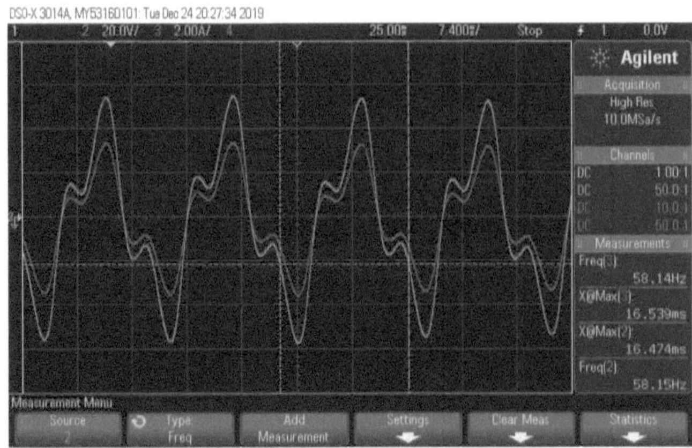

Figure 8-13: Phase voltage (green) and current (blue) waveforms using a capacitive load of 105nF

The harmonic content is evident in the voltage and current harmonic spectrum shown in Figure 8-14. The non-linear load did nothing to alter the harmonic content. The current THD is 11% while the voltage THD is a disturbing 40%.

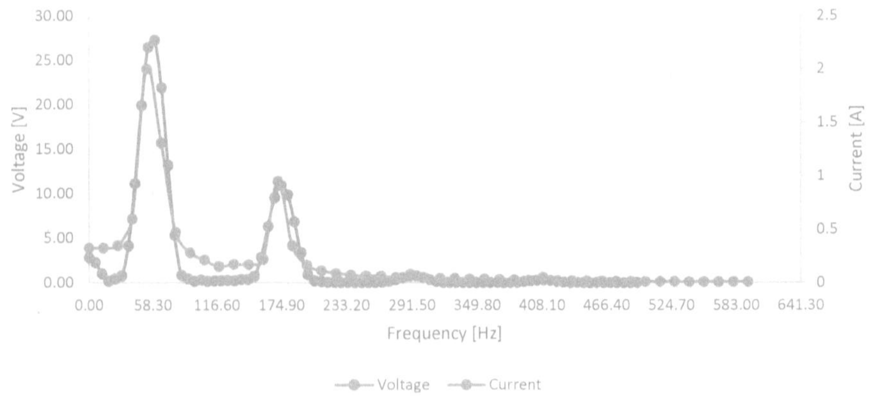

Figure 8-14: Harmonic spectrum of current and voltage under a capacitive load

Equation (3.11.2) is used to determine the output real power. The output apparent power is given by equation (3.6.1) and determined by the terminal voltage and load current. The power factor is calculated to be 0.74.

The Figure 6-14 in Section 6.3 can now be adapted to include the experimental results:

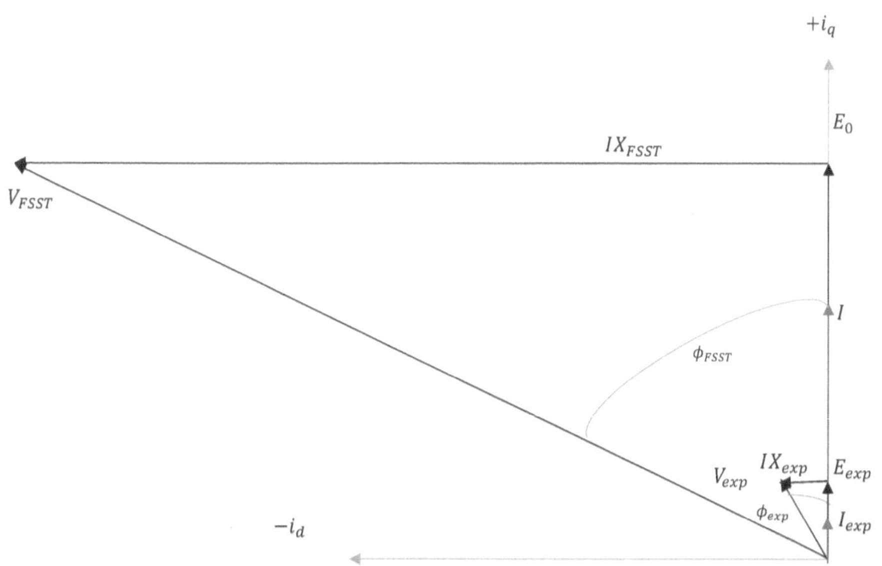

Figure 8-15: Phasor diagram of FEA prototype model and prototype

Figure 8-15 tells an interesting tale. As discussed previously, since the machine is operating with fewer turns per coil and lower load current than the FEA model, the voltage drop across the reactance is significantly smaller. To illustrate this, Figure 8-15 is drawn to scale to highlight that the experimental back-EMF is 5 times smaller than the designed, the load current is four times smaller, and the reactance is four times smaller. This is attributed to the fewer number of turns per coil and thus a lower inductance.

8.3.4 Efficiency

The efficiency of the prototype was determined by the ratio of the input shaft power to the output electrical power in equation (3.11.1). The calculation was performed over the speed range and the result is shown in Figure 8-16.

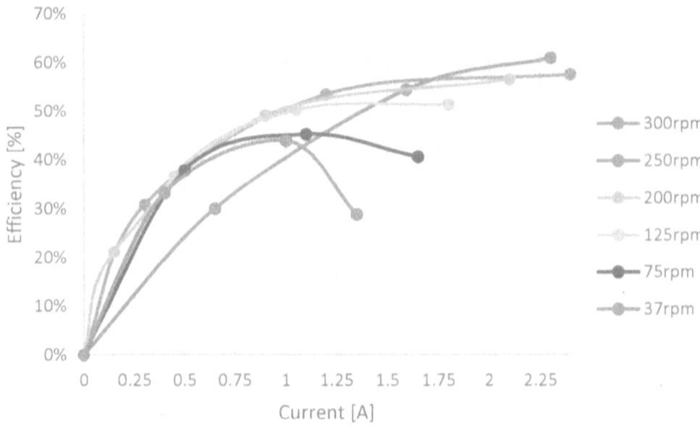

Figure 8-16: Efficiency vs load current at various shaft speeds

A purely resistive load was used across the test speed range, thus the prototype operated at unity power factor ($\cos(\theta) = 1$).

The efficiency of the prototype increases with load current until the point at which the copper losses overtake the core losses. This is attributed to the copper losses which increases with load current. The maximum efficiency of 61% is at 250RPM and 2.3A.

Additionally, the prototype displays lower efficiency at lower speed. This is attributed to the decrease in terminal voltage with decreasing speed.

Consequently, higher load current and lower speed yields the greatest drop in efficiency with the minimum efficiency of 29% at 37RPM and 1.3A. This can be attributed to the voltage drop across the stator winding resistance in comparison to the low terminal voltage.

8.3.5 Electrical Output Power

In addition to the efficiency, the output power over the speed range was also taken account of. The electrical output power with respect to load current at different shaft speed is shown in Figure 8-17.

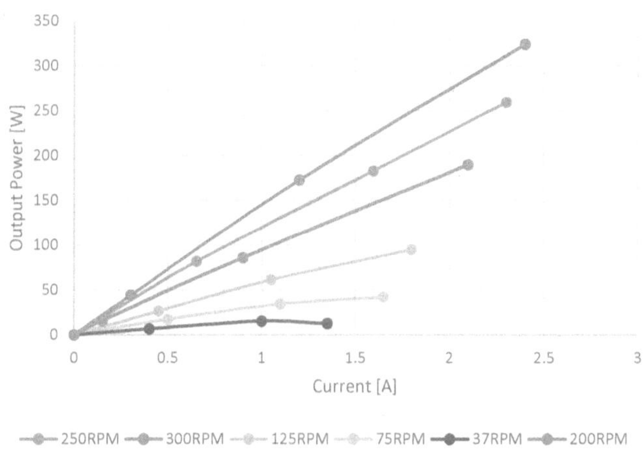

Figure 8-17: Electrical output power vs current for different speeds

Naturally, the lowest output power was at the lowest speeds. This would be due to the low terminal voltage at low speed.

9. Conclusion

This book assessed the suitability of a Vernier permanent magnet machine for a wind application. This was accomplished through the design and analysis of a fractional slot spoke type Vernier machine. Finally, the design was assessed using FEA and prototyped. The prototype was tested in the lab. The experiment results pointed to a manufacturing flaw which was determined as being mechanically deformed lamination material and fewer turns in the winding than was designed. The following sections shed further light on the study.

9.1 Literature

A literature survey was conducted on the history of the Vernier machine from its humble beginnings as a variable reluctance machine to modern variations such as the HTS superconducting Vernier machine. The various Vernier sub-topologies were presented. The Vernier machine's cousin, the integrated magnetic gear machine, was also briefly discussed. Research indicates that the magnetic gearing effect is being assessed in a wide variety of machine topologies. It seems to have peaked the interest of researchers owing to its low speed and high torque offering – highly desirable characteristics for wind turbines and electric vehicles. The research consensus was largely apparent: the Vernier topology offers low speed and high torque with one draw-back – low power factor. The dual stator Vernier topology has gained favour due to its mitigation of leakage flux which improves power factor. However, there were a few outliers who suggested that Vernier machines with spoke magnets and fractional slot windings can produce respectable power factor performance. This paper sought to explore such a machine. The results were not as desired, yet the gaps in Vernier knowledge have been highlighted. It can be concluded that these gaps can only point future Vernier researchers in the right direction.

The literature on the Vernier permanent magnet machine topology speaks of a lot of promise. Whether this topology will reach those heights remains to be seen. For now, the Vernier topology is still very much a niche topology.

9.2 Topology

The fractional slot spoke type Vernier machine was chosen. This was based on a 2D FEA comparative study of four Vernier topologies, namely; integral slot surface mounted (ISSM), integral slot spoke type (ISST), fractional slot surface mounted (FSSM), and finally the fractional slot spoke type (FSST). The FEA results indicated that the surface mounted topologies corresponded with the Vernier analytical more than the spoke type topologies. The spoke topologies showed a diminished Vernier magnetic flux density component in the harmonic spectra. This was predicted in literature. The spoke topologies thus required a bit more fine-tuning – the integral slot topology more so than the fractional slot topology. In fact, the ISST topology required greater magnet volume and windings to meet the performance criteria. This was not the case for the FSST topology which require relatively small increases in active material. Moreover, the FSST topology displayed the best power factor performance which was finally the deciding factor.

9.3 Sizing

Initially, this paper set out to design a 6kW Vernier PM machine for a rated speed of 250RPM. The Vernier machine analytical sizing approach (Section) in this paper has significant similarities to that of conventional PM machines. The only difference between the two approaches is the magnetic loading. The Vernier magnetic loading constitutes the Vernier component, the additional flux density component due to the slotting effect – a unique feature in Vernier machines. The prevalence of this component was evident throughout the prototype design phases. This gave validation to literature and the analytical analysis. However, it was not evident in the experimental results.

9.4 Numerical Analysis – FEA

In order to validate the analytical design and sizing, a finite element analysis was conducted. The 2D FEA results were the basis on which the comparative study in section was decided. The FSST topology was selected and subsequently studied in further detail. The design torque and load angle was analysed in depth using 2D FEA. The power factor and desired operating point of the machine in motoring mode was thus defined. The results of the study were validated using 3D FEA.

Moreover, 2D FEA was used to refine the structural integrity of the machine. 3D FEA was not used as the dimensionally small structural features created an excessively dense FEA mesh. The structural features were therefore over-designed, and the performance criteria were re-checked and compared to the original design. The structurally sound rotor showed similar performance characteristic with %% drop in performance torque, voltage, flux density.

9.5 Prototyping

The material was sourced from, and the prototype was manufactured, in Zhicheng, China. The windings were completed in Cape Town, South Africa.

The prototype was constructed based on FEA results which indicated favourable performance. The FEA model was refined and made even more practically realisable by improving the mechanical integrity of the design and assessing the effect the structural features had on electromagnetic performance. Due to manufacturing defects, the prototype fell short of the wind application criteria.

9.6 Wind generator suitability

The analytical and FEA results showed good correlation. This was deemed sufficient evidence to proceed with the manufacturing of the prototype.

Unfortunately, due to manufacturing defects, the prototype did not meet the desired criteria. The prototype fell short in all wind application criteria. It was seen in Section 8.2.1 that the back-EMF, and therefore the generated power, was less than designed. It follows that the Vernier contribution described in Section 3.3 was not present in the prototype.

Moreover, the phase back-EMF THD was larger than anticipated. Experimental results did not agree with the conclusions described in Section 5.1. The line back-EMF was sinusoidal and possessed relatively low THD of 10%, but the phase back-EMF had a THD of 35%. The high phase back-EMF THD was due to the significant 3rd harmonic component. A problematic 3rd harmonic in fractional slot Vernier machines, in spite of the gearing ratio, has not been analysed in literature sufficient enough as yet. It is hoped this book will pave the way for this.

Furthermore, the cogging torque and torque ripple was higher than expected. Here again, the effect of the gearing ratio, was not evident. Literature believed the presence of the gearing ratio brought about low torque harmonics. This was not the case.

Ultimately the poor performance can be attributed to three factors:

1) Machining of the stator bore which lead to break down and meshing of the lamination material - this undoubtedly resulted in additional eddy current loss which was not present in the FEA analysis
2) Winding defect - the back-EMF, resistance and inductance was much lower than FEA predicted
3) Deformation of the rotor during the initial testing which resulted from the rotor being in contact with stator while rotating - created an eccentric and irregular air-gap which undoubtedly exacerbated torque ripple

10. Recommendations

10.1 Literature

Further research into Vernier operating principle is needed. Particularly, the analytical description of the spoke type PM performance parameters. It is recommended that the next Vernier machine book focuses on developing the governing analytical theory rather than a prototype.

10.2 Topology

Future research should shed more light on how the FSST topology is absolved from having a low power factor despite its low pole ratio. The FEA model in this book as well as the prototype in [88] produced respectable power factors despite its comparatively small pole ratio.

10.3 Sizing

An optimization algorithm can be employed during the design phase. This can further improve efficiency and reduce dependence on the Vernier analytical theory which has been shown to be lacking.

10.4 Numerical Analysis

The accuracy of the FEA phase can be greatly improved by doing detailed 3D FEA analysis. This paper showed that there was favourable agreement between analytical and FEA results in the case of the surface mounted PM designs. There was some uncertainty in the spoke type PM designs. It is believed that the consequences of these uncertainties in the prototyping phase can be mitigated by analysing the spoke type design in 3D as well as 2D. The 3D model would take account of end effects and eddy effects in the surrounding steel. The cost of 3D FEA analyses is computing power. At the time of writing this book, there was no compatibility between the FEA software used and the UCT High Performance Computing Cluster.

10.5 Prototyping

Take control of prototype material quality during procurement and manufacturing stages. This can be accomplished by implementing factory acceptance tests of components. The factory tests will ensure that manufacturers comply with the relevant standards:

a. Electrical steel laminations
b. copper windings
c. permanent magnets
d. electrical machines

This will ensure all material and thus the final prototype meets quality and safety requirements.

Implement Hall sensors to measure the air-gap flux density. This will allow for the generating of the flux density harmonic spectrum in which the Vernier component can be experimentally analysed.

A less crude harmonic mitigation strategy would be to implement an LCL filter on the grid-side. The LCL filter enables the use of a voltage source converter (VSC) as an active rectifier. The VSC introduces current harmonics due its high switching frequencies but offers full control of the DC-link voltage, and even the power factor. A more cost-effective harmonic attenuation option is provided by the LCL filter [65].

References

[1] Z. Chen, J. M. Guerrero and F. Blaabjerg, "A Review of the State of the Art of Power Electronics for Wind Turbines," *IEEE Transactions on Power Electronics,* vol. 24, no. 8, pp. 1859-1875, 2009.

[2] M. Molina and P. Mercado, "Modelling and Control Design of Pitch-Controlled Variable Speed Wind Turbines, Wind Turbines, Dr. Ibrahim Al-Bahadly (Ed.), InTech," 4 April 2011. [Online]. Available: http://www.intechopen.com/books/wind-turbines/modelling-and-control-design-of-pitch-controlled-variable-speed-wind-turbines. [Accessed 2 February 2017].

[3] H. Polinder, F. F. van der Pijl and G.-J. T. P. J. de Vilder, "Comparison of Direct-Drive and Geared Generator Concepts for Wind Turbines," *IEEE Transactions on Energy Conversion,* vol. 21, no. 3, pp. 725-733, 2006.

[4] F. Blaabjerg and K. Ma, "Future on Power Electronics for Wind Turbine Systems," *IEEE Journal of Emerging and Selected Topics in POwer Electronics,* vol. 1, no. 3, pp. 139-152, 2013.

[5] K. J. Strnat, "The Recent Development of Permanent Magnet Materials Containing Rare Earth Metals," *IEEE Transactions on Magnetics,* vol. 6, no. 2, pp. 182-190, 1970.

[6] T. Burton, D. Sharpe, N. Jenkins and E. Bossanyi, "The Betz Limit," in *Wind Energy Handbook*, Chichester, John Wiley & Sons, LTD, 2001, pp. 45-46.

[7] A. Rashwan, M. Sayed, Y. Mobarak, G. Shabib and G. Buja, "Power Transition Enhancement for Variable-Speed, Variable-Pitch Wind Turbines using Model Predictive Control Techniques," in *2014 49th International Universities Power Engineering Conference (UPEC)*, Cluj-Napoca, 2014.

[8] J. Wanjiku, H. Jagau, M. Khan and P. Barendse, "Minimization of Cogging Torque in a Small Axial-Flux PMSG with a Parallel-teeth Stator," in *IEEE Energy Conversion Congress and Exposition*, Phoenix, 2011.

[9] P. Sen, "Voltage Regulation," in *Principles of Electric Machines and Power Electronics*, USA, John Wiley & Sons, Inc., 1997, pp. 58-62.

[10] X. Li, K. Chau and M. Cheng, "Analysis, Design and Experimental Verification of a Field-Modulated Permanent-Magnet Machine for Direct-Drive Wind Turbines," *IET Electric Power Applications,* vol. 9, no. 2, pp. 150-159, 2015.

[11] A. Toba and T. Lipo, "Generic Torque-Maximizing Design Methodology of Surface Permanent-Magnet Vernier Machine," *IEEE Transactions on Industry Applications,* vol. 36, no. 6, pp. 1539-1546, 2000.

[12] H. Yang, H. Lin, Z. Zhu, S. Fang and Y. Huang, "Novel Flux-Regulatable Dual-Magnet Vernier Memory Machines for Electric Vehicle Propulsion," *IEEE Transactions on Applied Superconductivity,* vol. 24, no. 5, p. 0601205, 2014.

[13] J. Li, K. Chau, J. Jiang, C. Liu and W. Li, "A New Efficient Permanent-Magnet Vernier Machine for Wind Power Generation," *IEEE Transactions on Magnetics,* vol. 46, no. 6, pp. 1475-1478, 2010.

[14] B. Kim and T. Lipo, "Operation and Design Principles of a PM Vernier Motor," *IEEE Transactions on Industry Applications,* vol. 50, no. 6, pp. 3656-3663, 2014.

[15] D. Li, R. Qu, J. Li, L. Xiao, L. Wu and W. Xu, "Analysis of Torque Capability and Quality in Vernier Permanent-Magnet Machines," *IEEE Transactions on Industry Applications,* vol. 52, no. 1, pp. 125-135, 2016.

[16] D. Li and Q. Ronghai, "Sinusoidal Back-EMF of Vernier Permanent Magnet Machines," in *Electrical Machines and Systems (ICEMS)*, Sapporo, 2012.

[17] K. Atallah and D. Howe, "A Novel High-Performance Magnetic Gear," *IEEE Transactions on Magnetics,* vol. 37, no. 4, pp. 2844-2846, 2001.

[18] L. Jian, K. Chau, D. Zhang, J. Jiang and Z. Wang, "A Magnetic-geared Outer-rotor Permanent-magnet Brushless Machine for Wind Power Generation," in *Industry Applications Conference* , New Orleans, 2007.

[19] L. Bronn, "Design and Performance Evaluation of a Magnetically Geared Axial-Flux Permanent Magnet Generator," Stellenbosch, Cape Town, 2012.

[20] S. Niu, S. Ho, W. Fu and L. Wang, "Quantitative Comparison of Novel Vernier Permanent Magnet Machines," *IEEE Transactions on Magnetics,* vol. 46, no. 6, pp. 2032-2035, 2010.

[21] M. Harris, S. Pajooman and S. Abu Sharkh, "The Problem of Power Factor in VRPM (Transverse-Flux) Machines," in *Electrical Machines and Drives*, Cambridge, 1997.

[22] D. Li, R. Qu and T. Lipo, "High Power Factor Vernier Permanent Magnet Machines," in *IEEE Energy Conversion Congress and Exposition*, Denver, 2013.

[23] B. Kim and T. Lipo, "Analysis of a PM Vernier Motor with Spoke Structure," *IEEE Transactions on Industry Applications,* vol. 52, no. 1, pp. 217-225, 2016.

[24] D. Li, R. Qu, W. Xu, J. Li and T. Lipo, "Design Procedure of Dual-stator, Spoke-Array Vernier Permanent Magnet Machines," *IEEE Transactions on Industry Applications,* vol. 51, no. 4, pp. 2972-2983, 2015.

[25] O. Dicke, "Vernier Motor". United States of America Patent 2,066,965, 23 May 1929.

[26] C. Lee, "Vernier Motor and Its Design," *IEEE Transactions on Power Apparatus and Systems,* vol. 82, no. 66, pp. 343-349, 1963.

[27] K. Mukherji and A. Tustin, "Vernier reluctance motor," *Proceedings of the Institution of Electrical Engineers,* vol. 121, no. 9, pp. 965-974, 1974.

[28] D. Rhodes, "Assessment of Vernier Motor Design using Generalised Machine Concepts," *IEEE Transactions on Power Apparatus and Systems ,* vol. 96, no. 4, pp. 1346-1352, 1977.

[29] A. Ishizaki, T. Tanka, K. Takasaki and S. Nishikata, "Theory and Optimum Design of PM Vernier Motor," in *Seventh International Conference on Electrical Machines and Drives, 1995,* Durham, 1995.

[30] A. Toba and T. Lipo, "Generic Torque-Maximizing Design Methodology of Surface Permanent-Magnet Vernier Machine," *IEEE Transactions on Industry Applications,* vol. 36, no. 6, pp. 1539-1546, 2000.

[31] B. Kim and T. Lipo, "Design of a Surface PM Vernier Motor for a Practical Variable Speed Application," in *2015 IEEE Energy Conversion Congress and Exposition (ECCE)*, Montreal, 2015.

[32] L. Wu, R. Qu, D. Li and Y. Gao, "Influence of Pole Ratio and Winding Pole Numbers on Performances and Optimal Design Parameters of Surface Permanent Magnet Vernier Machines," *IEEE Transactions on Industry Applications,* vol. 51, no. 5, pp. 3707 - 3715, 2015.

[33] D. Li, R. Qu and T. Lipo, "High-Power-Factor Vernier Permanent-Magnet Machines," *IEEE Transactions on Industry Applications,* vol. 50, no. 6, pp. 3664-3674, 2014.

[34] D. Li, R. Qu and J. Li, "Development and Experimental Evaluation of a Single-Winding, Dual-Stator, Spoke-Array Vernier Permanent Magnet Machines," in *IEEE Energy Conversion Congress and Exposition (ECCE)*, Montreal, 2015.

[35] Z. S. Du and T. A. Lipo, "Torque Performance Comparison Between a Ferrite Magnet Vernier Motor and an Industrial Interior Permanent Magnet Machine," *IEEE Transactions on Industry Applications,* vol. 53, no. 3, pp. 2088-2097, 2017.

[36] L. Sun and M. Cheng, "Improvement of Pole-Splitting Permanent-Magnet Vernier Machine with Permanent Magnets on both Stator and Rotor Sides," in *International Conference on Electrical Machines and Systems*, Hangzhou, 2014.

[37] W. Zhao, X. Sun, J. Ji and G. Liu, "Design and Analysis of New Vernier Permanent-Magnet Machine With Improved Torque Capability," *IEEE Transactions on Applied Superconductivity,* vol. 26, no. 4, p. 5201505, 2016.

[38] Y. Gao, R. Qu, J. Li, Z. Zhu and D. Li, "HTS Vernier Machine for Direct-Drive Wind Power Generation," *IEEE Transactions on Applied Superconductivity,* vol. 24, no. 5, p. 5202905, 2014.

[39] C. Cui, K. Li, X. Hu, H. Wang, Q. Wang and S. Zhao, "Analysis of Driving Torque Generated by Superconducting Motor Based on the Meissner Effect," in *Proceedings of 2015 IEEE International Conference on Applied Superconductivity and Electromagnetic Devices*, Shanghai, 2015.

[40] Y. Gao, R. Qu, D. Li, J. Li and G. Zhou, "Design of a Dual-Stator LTS Vernier Machine for Direct-Drive Wind Power Generation," *IEEE TRANSACTIONS ON APPLIED SUPERCONDUCTIVITY,* vol. 26, no. 4, p. 5204505, 2016.

[41] M. Shang, Y. Dai, Q. Wang, Y. Yu, B. Zhao, K. Kim and S. Oh, "Design and Electromagnetic Analysis of a Superconducting Diamagnetic Motor," *IEEE Transactions on Applied Superconductivity,* vol. 16, no. 2, pp. 1481-1484, 2006.

[42] P. Budig, "The Application of Linear Motors," in *Power Electronics and Motion Control Conference*, Beijing, 2000.

[43] Y. Kosuge, Y. Kataoka and M. Takayama, "Development of Surface Permanent Magnet-type Linear Vernier Motor," in *International Conference on Electrical Machines and Systems*, Pattaya City, 2015.

[44] Y. Du, M. Cheng, K. Chau, X. Liu, F. Xiao, W. Zhao, K. Shi and L. Mo, "Comparison of Linear Primary Permanent Magnet Vernier Machine and Linear Vernier Hybrid Machine," *IEEE Transactions on Magnetics,* vol. 50, no. 11, p. 8202604, 2014.

[45] S. Ho, S. Niu and W. Fu, "Design and Control of a Novel Axial Flux Design and Control of a Novel Axial Flux Permanent Magnet In-Wheel Machine for HEVs," 2011.

[46] X. Luo, S. Niu and W. Fu, "Design and Sensorless Control of a Novel Axial-Flux Permanent Magnet Machine for In-Wheel Applications," *IEEE Transactions on Applied Superconductivity,* vol. PP, no. 99, p. 1, 2016.

[47] T. Zou, R. Qu, J. Li and D. Li, "Analysis and Design of a Dual-Rotor Axial-Flux Vernier Permanent Magnet Machine," in *2015 IEEE Energy Conversion Congress and Exposition (ECCE)*, Montreal, 2015.

[48] F. Zhao, T. Lipo and B. Kwon, "A Novel Dual-Stator Axial-Flux Spoke-Type Permanent Magnet Vernier Machine for Direct-Drive Applications," *IEEE Transactions on Magnetics,* vol. 50, no. 11, p. 8104304, 2014.

[49] A. Toba and T. Lipo, "Novel Dual-Excitation Permanent Magnet Vernier Machine," in *Conference Record of the 1999 IEEE Industry Applications Conference, 1999. Thirty-Fourth IAS Annual Meeting*, Phoenix, 1999.

[50] S. Niu, S. Ho and W. Fu, "A Novel Direct-Drive Dual-Structure Permanent Magnet Machine," *IEEE Transactions on Magnetics,* vol. 46, no. 6, pp. 2036-2039, 2010.

[51] Y. Alamoudi, G. Atkinson, B. Mecrow and M. Zhang, "A New High Torque Density Permanent Magnet Machine Design for Electric Vehicles," in *IECON 2012 - 38th Annual Conference on IEEE Industrial Electronics Society*, Montreal, 2012.

[52] M. Takano and S. Shimomura, "Study of Variable Reluctance Vernier Motor for Hybrid Electric Vehicle," in *2013 IEEE ECCE Asia Downunder (ECCE Asia)* , Melbourne, 2013.

[53] G. Liu, J. Yang, W. Zhao, J. Ji, Q. Chen and W. Gong, "Design and Analysis of a New Fault-Tolerant Permanent-Magnet Vernier Machine for Electric Vehicles," *IEEE Transactions on Magnetics,* vol. 48, no. 11, pp. 4176-4179, 2012.

[54] D. Matt, J. Jac and N. Ziegler, "Design of a Mean Power Wind Conversion Chain with a Magnetic Speed Multiplier," 21 November 2012. [Online]. Available: http://www.intechopen.com/books/advances-in-wind-power/design-of-a-mean-power-wind-conversion-chain-with-a-magnetic-speed-multiplier. [Accessed 19 January 2017].

[55] Z. Zhu and D. Howe, "Instantaneous Magnetic Field Distribution in Brushless Permanent Magnet DC Motors, Part III: Effect of Stator Slotting," *IEEE Transactions on Magnetics,* vol. 29, no. 1, pp. 143-151, 1993.

[56] J. Hsu, B. Scoggins, M. Scudiere, L. Marlino, D. Adams and P. Pillay, "Nature and Assessments of Torque Ripples of Permanent-Magnet Adjustable-Speed Motors," in *Industry Applications Conference, 1995. Thirtieth IAS Annual Meeting, IAS '95., Conference Record of the 1995 IEEE,* Orlando, 1995.

[57] N. Bianchi and M. Dai Pre, "Use of the star of slots in designing fractional-slot single-layer synchronous motors," *IEE Proceedings - Electric Power Applications,* vol. 153, no. 3, pp. 459 - 466, 2006.

[58] J. Hendershot and T. Miller, "Winding Inductances and Armature Reaction," in *Design of Brushless Permanent-Magnet Motors,* USA, Magna Physics Publishing, 1994, pp. 5-39 - 5-59.

[59] P. Sen, "Inductance," in *Principles of Electric Machines and Power Electronics*, Chennai, India, Wiley, 2013, pp. 13-14.

[60] J. Kirtley, "6.685 Electric Machines," in *Class Notes 7: Permanent Magnet "Brushless DC" Motors*, Massachusetts, Massachusetts Institute of Technology - Department of Electrical Engineering and Computer Science, 2005.

[61] D. Hanselman, "Coil Inductance," in *Brushless Permanent Magnet Motor Design*, Lebanon, Ohio, Magna Physics Publisher, 2006, pp. 94-98.

[62] P. Sen, "Balanced Three-Phase Circuits," in *Principles of Electric Machines and Power Electronics*, USA, John Wiley & Sons, 1997, pp. 583-600.

[63] J. De Kooning, B. Meersman, T. Vandoorn and L. Vandevelde, "Efficiency Improvement of a Small Wind Turbine by Adaptation of the Rectifier Circuit and Control," in *Researchers Symposium in Electrical Power Engineering, Proceedings*, Netherlands, 2012.

[64] J. De Kooning, J. Van de Vywer, T. Vandoorn, B. Meersman and L. Vandevelde, "Joule Losses and Torque Ripple Caused by Current Waveforms in Small and Medium Wind Turbines," in *EUROCON*, Zagreb, 2013.

[65] M. Liserre, F. Blaabjerg and S. Hansen, "Design and Control of an LCL-filter based Three-phase Active Rectifier," in *Industry Applications Conference*, Chicago, 2001.

[66] C. da Silva, F. Bidaud, P. Herbert and J. Cardoso, "Power Factor Calculation by the Finite Element Method," *IEEE Transactions on Magnetics,* vol. 46, no. 8, pp. 3002-3005, 2010.

[67] W. Grady and R. Gilleskie, "Harmonics and How They Relate to Power Factor," in *EPRI Power Quality & Opportunities Conference*, San Diego, 1993.

[68] Z. Zhu and D. Howe, "Influence of Design Parameters on Cogging Torque in Permanent Magnet Machines," *IEEE Transactions on Energy Conversion,* vol. 15, no. 4, pp. 407-412, 2000.

[69] D. Lin, P. Zhou, C. Lu and S. Lin, "Analytical Prediction of Cogging Torque for Spoke Type Permanent Magnet Machines," *IEEE Transactions on Magnetics,* vol. 48, no. 2, pp. 1035-1038, 2012.

[70] D. Hanselman, "Cogging Torque," in *Brushless Permanent Magnet Motor Design - Second Edition*, Cranston, Rhode Island, Magna Physics Publishing, 2006, pp. 111-115.

[71] L. Dosiek and P. Pillay, "Cogging Torque Reduction in Permanent Magnet Machines," *IEEE Transactions on Industry Applications,* vol. 43, no. 6, pp. 1565-1571, 2007.

[72] D. Hanselman, "Performance," in *Brushless Permanent Magnet Motor Design*, Lebanon, Ohio, Magna Physics Publishing, 2006, pp. 203 - 226.

[73] J. Gieras and M. Wing, "Winding Losses," in *Permanent Magnet Motor Technology Design and Applications*, New York, USA, Marcel Dekker, Inc, 2002, pp. 243 - 245.

[74] O. Aglen and A. Andersson, "Thermal analysis of a high-speed generator," in *38th IAS Annual Meeting on Conference Record of the Industry Applications Conference*, Salt Lake City, 2003.

[75] P. Sen, "Core Loss," in *Principles of Electric Machines and Power Electronics*, Chennai, India, Wiley, 2014, pp. 20 - 21.

[76] R. Dutta and M. Rahman, "Comparison of Core Loss Prediction Methods for the Interior Permanent Magnet Machine," in *2005 International Conference on Power Electronics and Drives Systems*, Kuala Lumpur, Malaysia, 2005.

[77] J. Gieras, R. Wang and M. Kamper, Axial Flux Permanent Magnet Brushless Machines, 2 ed., Springer, 2008.

[78] J. Gieras and M. Wing, "Rotational Losses," in *Permanent Magnet Motor Technology*, New York, Marcel Dekker Inc, 2002, pp. 555-556.

[79] L. Xu, G. Liu, W. Zhao, j. Ji, H. Zhou, W. Zhao and T. Jiang, "Quantitative Comparison of Integral and Fractional Slot Permanent Magnet Vernier Motors," *IEEE Transactions on Energy Conversion* , vol. 30, no. 4, pp. 1483 - 1495, 2015.

[80] X. Li, K. T. Chau and M. Cheng, "Comparative Analysis and Experimental Verification of an Effective Permanent-Magnet Vernier Machine," *IEEE Transactions on Magnetics,* vol. 51, no. 7, p. 8203009, 2015.

[81] D. Hanselman, "General Sizing," in *Brushless Permanent Magnet Motor Design Second Edition*, Lebanon, Ohio, Magna Physics Publishing, 2006, pp. 203-206.

[82] V. S. Division, *Data Sheet Isovac 530-50A Electrical Steel,* Linz, Austria: Voestalpine Steel Division, 2018.

[83] J. Hendershot and T. Miller, "Basic Design Choices," in *Design of Brushless Permanent-Magnet Motors*, Hillsboro, Ohio, Magna Physics Publishing, 1994.

[84] B.-K. Lee, G.-H. Kang, J. Hur and D.-W. You, "Design of Spoke Type BLDC Motors with High," *Industry Applications Conference, 2004. 39th IAS Annual Meeting. Conference Record of the 2004 IEEE,* vol. 2, pp. 1068 - 1074, 2004.

[85] G. Dajaku and D. Gerling, "Magnetic Radial Force Density of the PM Machine with 12-teeth/10-poles winding topology," in *2009 IEEE International Electric Machines and Drives Conference*, Miami, USA, 2009.

[86] H. Jagau, *Design of a Permanent Magnet Generator for a Sustainable Wind Energy Capture and Storage System,* Cape Town: University of Cape Town, 2011.

[87] J. Wanjiku, *Design of an Axial-Flux Generator for a Small-Scale Wind Electrolyser Plant,* Cape Town: University of Cape Town, 2010.

[88] X. Li, K. Chau and M. Cheng, "Comparative Analysis and Experimental Verification of an Effective Permanent-Magnet Vernier Machine," *IEEE Transactions on Magnetics,* vol. 51, no. 7, pp. 8203009-8203009, 2015.

[89] D. Dudley, "Design of a Vernier Permanent Magnet Wind Generator," University of Cape Town, Cape Town, South Africa, 2020.

[90] L. Papini, C. Gerada and P. Bolognesi, "Comparison of Analytical and FE Modeling of Vernier Hybrid Machine," in *2012 XXth International Conference on Electrical Machines (ICEM)*, Marseille, 2012.

[91] J. Yang, G. liu, W. Zhao, Q. Chen, Y. Jiang, L. Sun and X. Zhu, "Quantitative Comparison for Fractional-Slot Concentrated-Winding Configurations of Permanent-Magnet Vernier Machines," *IEEE Transactions on Magnetics,* vol. 49, no. 7, pp. 3826-3829, 2013.

[92] K. Okada, N. Niguchi and K. Hirata, "Analysis of a Vernier Motor with Concentrated Windings," *IEEE Transactions on Magnetics,* vol. 49, no. 5, pp. 2241-2244, 2013.

[93] G. De Donato, F. Capponi, G. Rivellini and F. Caricchi, "Integral-Slot Versus Fractional-Slot Concentrated-Winding Axial-Flux Permanent-Magnet Machines: Comparative Design, FEA, and Experimental Tests," *IEEE Transactions on Industry Applications,* vol. 48, no. 5, pp. 1487-1495, 2012.

[94] J. Hendershot and T. Miller, "Sinewave Motors," in *Design of Brushless Permanent-Magnet Motors*, Oxford, Magna Physics Publishing and Clarendon Press, 1994, pp. 6-23 - 6-33.

[95] J. D. Kramer and C. Jacky, "How to write biblos," vol. 1, no. 1, 2006.

[96] H. Polinder, S. de Haan, M. Dubois and J. Slootweg, "Basic Operation Principles and Electrical Conversion Systems of Wind Turbines," *European Power Electronics and Drives,* vol. 15, no. 4, pp. 1-9, 2005.

[97] D. Staton, T. Miller and S. Wood, "Maximising the saliency ratio of the synchronous reluctance motor," *IEE Proceedings B - Electric Power Applications,* vol. 140, no. 4, pp. 249 - 259, 1993.

[98] P. Niazi, H. Toliyat, D. Cheong and J. Kim, "A Low-Cost and Efficient Permanent-Magnet-Assisted Synchronous Reluctance Motor Drive," *IEEE Transactions on Industry Applications,* vol. 43, no. 2, pp. 542-550, 2007.

[99] J. Haataja, "A Comparative Performance Study of Four-Pole Induction Motors and Synchronous Reluctance Motors in Variable Speed Drives," Lappeenranta University of Technology, Lappeenranta, 2003.

[100] T. Hubert, M. Reinlein, A. Kremser and H. Herzog, "Preliminary Design of Reluctance Synchronous Machines using simplified Magnetic Circuit Analysis," in *IEEE 5th Conference on Power Engineering, Energy and Electrical Drives*, Riga, 2015.

[101] T. Mohanarajah, J. Rizk, M. Nagrial and A. Hellany, "Design Optimisation of Flux Barrier Synchronous Reluctance Machines," in *2015 International Conference on Optimization of Electrical & Electronic Equipment*, Side, 2015.

[102] N. Ozcelik, U. Dogru, H. Gedik, M. Imeruz and L. Ergene, "A Multi-parameter Analysis for Rotor Design of Synchronous Reluctance Motors," in *2016 XXII International Conference on Electrical Machines (ICEM)*, Lausanne, 2016.

[103] E. Montalvo-Ortiz, S. Foster, J. Cintron-Rivera and E. Strangas, "Comparison between a spoke-type PMSM and a PMASynRM using ferrite magnets," in *2013 IEEE International Electric Machines and Drives Conference*, Chicago, 2013.

[104] T. Tokuda, M. Sanada and S. Morimoto, "Influence of Rotor Structure on Performance of Permanent Magnet Assisted Synchronous Reluctance Motor," in *International Conference on Electrical Machines and Systems, 2009*, Tokyo, 2009.

[105] A. Kronberg, *Design and Simulation of Field Oriented Control and Direct Torque Control for a Permanent Magnet Synchronous Motor with Positive Saliency,* Upsalla: Upsalla University, 2012.

[106] J. Kolzer, T. Bazzo, R. Carlson and F. Wurtz, "Comparative analysis of a spoke ferrite permanent magnets vernier synchronous generator," in *2016 XXII International Conference on Electrical Machines (ICEM)*, Lausanne, Switzerland, 2016.

[107] J. Gieras and W. Wing, "Stray Losses," in *Permanent Magnet Motor Technology*, New York, USA, Marcel Dekker, Inc, 2002, pp. 554 - 555.

Appendices

Appendix A Slotting Effect

The following analysis uses the conformal mapping method to model the variable air-gap permeance due to stator tooth slotting in a permanent magnet machine. In conventional machines the slotting effect is accounted for by the Carter's Coefficient.

Henceforth, the open-circuit flux density is calculated from the product of the field produced by the magnets when stator slotting is neglected and the relative permeance function of the slotted air-gap region.

The 2-D permeance function is introduced:

$$\lambda(\alpha, r) = \begin{cases} \Lambda_0 \left[1 - \beta(r) - \beta(r)\cos\left(\frac{\pi}{0.8\alpha_0}\alpha\right)\right] & ; for\ 0 \leq \alpha \leq 0.8\alpha_0 \\ \Lambda_0 & ; for\ 0.8 \leq \alpha \leq 0.5\alpha_t \end{cases} \tag{A. 1}$$

The parameter $\beta(r)$ is derived using the conformal transformation [55] and is described as:

$$\beta(r) = \frac{1}{2}\left[1 - \frac{1}{\sqrt{1 + \left(\frac{b_0}{2g'}\right)^2 (1 + v^2)}}\right] \tag{A. 2}$$

The relative permeance function is calculated from:

$$\begin{aligned} \tilde{\lambda}(\alpha, r) &= \frac{\lambda(\alpha, r)}{\Lambda_0} \\ &= \frac{\lambda(\alpha, r)}{\mu_0 / g'} \end{aligned} \tag{A. 3}$$

In its Fouier Series form, equation A.3 is represented as:

$$\begin{aligned} \tilde{\lambda}(\alpha, r) &= \sum_{\mu=0}^{\infty} \tilde{\Lambda}_\mu(r) \cos\{\mu Q(\alpha + \alpha_{sa})\} \\ &= \tilde{\Lambda}_0 + \sum_{\mu=1}^{\infty} \tilde{\Lambda}_\mu(r) \cos\{\mu Q(\alpha + \alpha_{sa})\} \end{aligned} \tag{A. 4}$$

The constant term in the Fourier series is $\tilde{\Lambda}_0$, it is described as:

$$\widetilde{\Lambda}_0 = \frac{2}{\alpha_t} \int_{-\frac{\alpha_t}{2}}^{\frac{\alpha_t}{2}} \tilde{\lambda}(\alpha, r) d\alpha \tag{A.5}$$

Substitute in equation A.1:

$$\widetilde{\Lambda}_0 = \frac{2}{\alpha_t} \int_0^{0.8\alpha_0} \left(1 - \beta - \beta \cos\left(\frac{\pi\alpha}{0.8\alpha_0}\right)\right) d\alpha + \frac{2}{\alpha_t} \int_{0.8\alpha_0}^{0.5\alpha_t} d\alpha \tag{A.6}$$

Expand brackets and integrals:

$$\widetilde{\Lambda}_0 = \frac{2}{\alpha_t} \left\{ \int_0^{0.8\alpha_0} d\alpha - \int_0^{0.8\alpha_0} \beta d\alpha - \int_0^{0.8\alpha_0} \beta \cos\left(\frac{\pi\alpha}{0.8\alpha_0}\right) d\alpha + \int_{0.8\alpha_0}^{0.5\alpha_t} d\alpha \right\} \tag{A.7}$$

Integrate:

$$\widetilde{\Lambda}_0 = \frac{2}{\alpha_t} \left\{ 0.8\alpha_0 - 0.8\alpha_0\beta - \beta \frac{0.8\alpha_0}{\pi} \left[\sin\left(\frac{\pi\alpha}{0.8\alpha_0}\right) \right]_0^{0.8\alpha_0} + (0.5\alpha_t - 0.8\alpha_0) \right\} \tag{A.8}$$

Simplify:

$$\widetilde{\Lambda}_0 = \frac{2}{\alpha_t} \left\{ -0.8\alpha_0\beta + 0.5\alpha_t - \beta \frac{0.8\alpha_0}{\pi} [\sin(\pi) - \sin(0)] \right\} \tag{A.9}$$

This results in the final expression for the constant Permeance coefficient:

$$\widetilde{\Lambda}_0 = \left\{ 1 - \frac{1.6\alpha_0}{\alpha_t} \beta \right\} \tag{A.10}$$

If you define: $\alpha_t = \frac{\tau}{R_s}$ *and* $\alpha_0 = \frac{b_0}{R_s}$ and recall that $\Lambda_0 = \widetilde{\Lambda}_0 \frac{\mu_0}{g'}$ then the result is the same constant permeance coefficient that is described in equation (3.1.3) in section 3.1:

$$\Lambda_0 = \frac{\mu_0}{g'} \left(1 - 1.6 \frac{b_0}{\tau} \beta \right) \tag{A.11}$$

The air-gap permeance harmonic component is calculated using Fourier series theory as:

$$\widetilde{\Lambda}_\mu = \frac{2}{\alpha_t} \int_{-\frac{\alpha_t}{2}}^{\frac{\alpha_t}{2}} \tilde{\lambda}(\alpha, r) \cos(kQ\alpha)\, d\alpha \tag{A.12}$$

Substitute equation A.1:

$$\widetilde{\Lambda}_{\mu} = \frac{4}{\alpha_t}\int_{0}^{0.8\alpha_0}\left(1-\beta-\beta\cos\left(\frac{\pi\alpha}{0.8\alpha_0}\right)\right)\cos(kQ\alpha)\,d\alpha + \frac{4}{\alpha_t}\int_{0.8\alpha_0}^{0.5\alpha_t}\cos(kQ\alpha)\,d\alpha \tag{A. 13}$$

Expand brackets and integrals:

$$\widetilde{\Lambda}_{\mu} = \frac{4}{\alpha_t}\left\{\int_{0}^{0.8\alpha_0}\cos(kQ\alpha)\,d\alpha - \int_{0}^{0.8\alpha_0}\beta\cos(kQ\alpha)\,d\alpha - \int_{0}^{0.8\alpha_0}\beta\cos\left(\frac{\pi\alpha}{0.8\alpha_0}\right)\cos(kQ\alpha)\,d\alpha + \int_{0.8\alpha_0}^{0.5\alpha_t}\cos(kQ\alpha)\,d\alpha\right\} \tag{A. 14}$$

Expand using trigonometric identities:

$$\widetilde{\Lambda}_{\mu} = \frac{4}{\alpha_t}\left\{\int_{0}^{0.8\alpha_0}\cos(kQ\alpha)\,d\alpha - \int_{0}^{0.8\alpha_0}\beta\cos(kQ\alpha)\,d\alpha - \frac{\beta}{2}\int_{0}^{0.8\alpha_0}\left[\cos\alpha\left(\frac{\pi}{0.8\alpha_0}-kQ\right)+\cos\alpha\left(kQ+\frac{\pi}{0.8\alpha_0}\right)\right]d\alpha + \int_{0.8\alpha_0}^{0.5\alpha_t}\cos(kQ\alpha)\,d\alpha\right\} \tag{A. 15}$$

Simplify:

$$\widetilde{\Lambda}_{\mu} = \frac{4}{\alpha_t}\left\{\frac{1}{kQ}\sin(0.8\alpha_0 kQ) - \frac{\beta}{kQ}\sin(0.8\alpha_0 kQ) - \frac{\beta}{2}\int_{0}^{0.8\alpha_0}\left[\cos\alpha\left(\frac{\pi-0.8\alpha_0 kQ}{0.8\alpha_0}\right)+\cos\alpha\left(\frac{0.8\alpha_0 kQ+\pi}{0.8\alpha_0}\right)\right]d\alpha + \frac{1}{kQ}\sin(0.5\alpha_t kQ) - \frac{1}{kQ}\sin(0.8\alpha_0 kQ)\right\} \tag{A. 16}$$

Now since α_t represents the stator tooth pitch, the number of slots is given by $Q = \frac{2\pi}{\alpha_t}$. Then:

$$\tilde{\Lambda}_\mu = \frac{4}{\alpha_t}\left\{-\frac{\beta}{k\,(2\pi/\alpha_t)}\sin\left(0.8\alpha_0 k\frac{2\pi}{\alpha_t}\right) - \frac{\beta}{2}\int_0^{0.8\alpha_0}\left[\cos\alpha\left(\frac{\pi - 0.8\alpha_0 k\,(2\pi/\alpha_t)}{0.8\alpha_0}\right) + \cos\alpha\left(\frac{0.8\alpha_0 k\,(2\pi/\alpha_t) + \pi}{0.8\alpha_0}\right)\right]d\alpha + \frac{1}{kQ}\sin\left(0.5\alpha_t k\frac{2\pi}{\alpha_t}\right)\right\} \tag{A. 17}$$

Simplify:

$$\tilde{\Lambda}_\mu = \frac{4}{\alpha_t}\left\{-\frac{\beta\alpha_t}{2\pi k}\sin\left(1.6\pi k\frac{b_0}{\tau}\right) - \frac{\beta}{2}\frac{0.8\alpha_0}{\pi - 1.6\pi k\frac{b_0}{\tau}}\left[\sin\alpha\left(\frac{\pi - 1.6\pi k\frac{b_0}{\tau}}{0.8\alpha_0}\right)\right]_0^{0.8\alpha_0} - \frac{\beta}{2}\frac{0.8\alpha_0}{\pi + 1.6\pi k\frac{b_0}{\tau}}\left[\sin\alpha\left(\frac{1.6\pi k\frac{b_0}{\tau} + \pi}{0.8\alpha_0}\right)\right]_0^{0.8\alpha_0}\right\} \tag{A. 18}$$

Simplify remaining integrands

$$\tilde{\Lambda}_\mu = \frac{4}{\alpha_t}\left\{-\frac{\beta\alpha_t}{2\pi k}\sin\left(1.6\pi k\frac{b_0}{\tau}\right) - \frac{\beta}{2}\frac{0.8\alpha_0}{\pi - 1.6\pi k\frac{b_0}{\tau}}\sin\left(1.6\pi k\frac{b_0}{\tau}\right) + \frac{\beta}{2}\frac{0.8\alpha_0}{\pi + 1.6\pi k\frac{b_0}{\tau}}\sin\left(1.6\pi k\frac{b_0}{\tau}\right)\right\} \tag{A. 19}$$

Take out common factors:

$$\tilde{\Lambda}_\mu = \frac{2\beta}{\alpha_t}\left\{-\frac{\alpha_t}{\pi k} - \frac{0.8\alpha_0}{\pi - 1.6\pi k\frac{b_0}{\tau}} + \frac{0.8\alpha_0}{\pi + 1.6\pi k\frac{b_0}{\tau}}\right\}\sin\left(1.6\pi k\frac{b_0}{\tau}\right) \tag{A. 20}$$

Rearrange:

$$\tilde{\Lambda}_\mu = \frac{2\beta}{\alpha_t}\left\{-\frac{\alpha_t}{\pi k} - \left[\frac{0.8\alpha_0\left(\pi + 1.6\pi k\frac{b_0}{\tau}\right) - 0.8\alpha_0\left(\pi - 1.6\pi k\frac{b_0}{\tau}\right)}{\left(\pi - 1.6\pi k\frac{b_0}{\tau}\right)\left(\pi + 1.6\pi k\frac{b_0}{\tau}\right)}\right]\right\}\sin\left(1.6\pi k\frac{b_0}{\tau}\right) \tag{A. 21}$$

The following steps simplify and then manipulate the numerical terms to give the expression in section 3.1.

$$\widetilde{\Lambda}_{\mu} = \frac{2\beta}{\alpha_t}\left\{-\frac{\alpha_t}{\pi k}-\left[\frac{\alpha_0(0.8)(1.6)(2)\left(\pi k\frac{b_0}{\tau}\right)}{\pi^2-(1.6\pi)^2\left(k\frac{b_0}{\tau}\right)^2}\right]\right\}\sin\left(1.6\pi k\frac{b_0}{\tau}\right) \tag{A. 22}$$

$$=\frac{2\beta}{1}\left\{-\frac{1}{\pi k}-\frac{\alpha_0}{\alpha_t}\left[\frac{(0.8)(0.8)(4)\left(\pi k\frac{b_0}{\tau}\right)}{\pi^2-(2)(0.8)^2(\pi)^2(2)\left(k\frac{b_0}{\tau}\right)^2}\right]\right\}\sin\left(1.6\pi k\frac{b_0}{\tau}\right)$$

$$=\frac{2\beta}{1}\left\{-\frac{1}{\pi k}-\frac{1}{\pi k}\left[\frac{(0.8)^2(4)\left(k\frac{b_0}{\tau}\right)^2}{1-(2)(0.8)^2(2)\left(k\frac{b_0}{\tau}\right)^2}\right]\right\}\sin\left(1.6\pi k\frac{b_0}{\tau}\right)$$

$$=\frac{2\beta}{1}\left\{-\frac{1}{\pi k}-\frac{1}{\pi k}\left[\frac{(0.8)^2(4)\left(k\frac{b_0}{\tau}\right)^2}{(2)(0.8)^2\left(0.78125-2\left(k\frac{b_0}{\tau}\right)^2\right)}\right]\right\}\sin\left(1.6\pi k\frac{b_0}{\tau}\right)$$

$$=\frac{2\beta}{1}\left\{-\frac{1}{\pi k}-\frac{2}{\pi k}\left[\frac{\left(k\frac{b_0}{\tau}\right)^2}{0.78125-2\left(k\frac{b_0}{\tau}\right)^2}\right]\right\}\sin\left(1.6\pi k\frac{b_0}{\tau}\right)$$

$$=-\frac{4\beta}{\pi k}\left\{\frac{1}{2}+\frac{\left(k\frac{b_0}{\tau}\right)^2}{0.78125-2\left(k\frac{b_0}{\tau}\right)^2}\right\}\sin\left(1.6\pi k\frac{b_0}{\tau}\right)$$

Thus, the fundamental air-gap permeance component is given by:

$$\Lambda_1=-\frac{\mu_0}{g'}\frac{4}{\pi}\beta\left\{\frac{1}{2}+\frac{\left(\frac{b_0}{\tau}\right)^2}{0.78125-2\left(\frac{b_0}{\tau}\right)^2}\right\}\sin\left(1.6\pi k\frac{b_0}{\tau}\right) \tag{A. 23}$$

Appendix B   Pole-Pair Selection Tables

Table B. 1: Pole-dependent parameters for 1 winding pole-pair

p_r	Z	q	Pole-ratio	LCM	Cogging factor	Winding factor
5	6	1	5	30	2	1
8	9	1.5	8	144	1	0.945
11	12	2	11	132	2	0.966
14	15	2.5	14	420	1	0.951
17	18	3	17	306	2	0.96
20	21	3.5	20	840	1	0.953
23	24	4	23	552	2	0.958
26	27	4.5	26	1404	1	0.954
29	30	5	29	870	2	0.957
32	33	5.5	32	2112	1	0.954
35	36	6	35	1260	2	0.956

Table B. 2: Pole-dependent parameters for 2 winding pole-pairs

p_r	Z	q	Pole-ratio	LCM	Cogging factor	Winding factor
4	6	0.5	2	24	2	0.866
7	9	0.75	3.5	126	1	0.945
10	12	1	5	60	4	1
13	15	1.25	6.5	390	1	0.91
16	18	1.5	8	288	2	0.945
19	21	1.75	9.5	798	1	0.953
22	24	2	11	264	4	0.966
25	27	2.25	12.5	1350	1	0.941
28	30	2.5	14	840	2	0.951
31	33	2.75	15.5	2046	1	0.954
34	36	3	17	612	4	0.96

Table B. 3: Pole-dependent parameters for 3 winding pole-pairs

p_r	Z	q	Pole-ratio	LCM	Cogging factor	Winding factor
3	6	0.333	1	6	6	1
6	9	0.5	2	36	3	0.866
9	12	0.667	3	36	6	0.966
12	15	0.833	4	120	3	0.91
15	18	1	5	90	6	1
18	21	1.167	6	252	3	0.932
21	24	1.333	7	168	6	0.958
24	27	1.5	8	432	3	0.945
27	30	1.667	9	270	6	0.957
30	33	1.833	10	660	3	0.946
33	36	2	11	396	6	0.966

Table B. 4: Pole-dependent parameters for 4 winding pole-pairs

p_r	Z	q	Pole-ratio	LCM	Cogging factor	Winding factor
2	6	0.25	0.5	12	2	0
5	9	0.375	1.25	90	1	0.945
8	12	0.5	2	48	4	0.866
11	15	0.625	2.75	330	1	0.711
14	18	0.75	3.5	252	2	0.945
17	21	0.875	4.25	714	1	0.89
20	24	1	5	120	8	1
23	27	1.125	5.75	1242	1	0.941
26	30	1.25	6.5	780	2	0.91
29	33	1.375	7.25	1914	1	0.954
32	36	1.5	8	576	4	0.945

Appendix C Bearing Datasheet

3207 A

Popular item
SKF Explorer

Dimensions

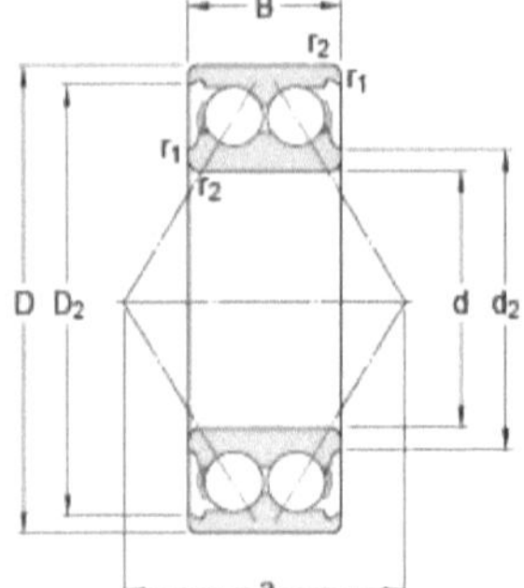

d		35	mm
D		72	mm
B		27	mm
d_2	≈	45.4	mm
D_2	≈	63.85	mm
$r_{1,2}$	min.	1.1	mm
a		42	mm

Abutment dimensions

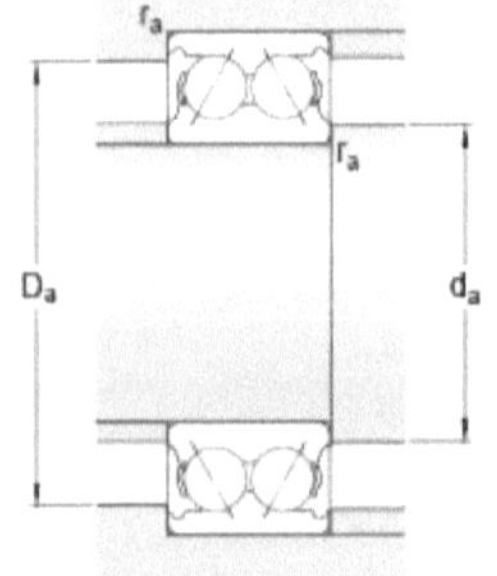

d_a	min.	42	mm
D_a	max.	65	mm
r_a	max.	1	mm

Calculation data

Basic dynamic load rating	C	40.5	kN
Basic static load rating	C_0	30	kN
Fatigue load limit	P_u	1.27	kN
Reference speed		9000	r/min
Limiting speed		9000	r/min
Calculation factor	k_r	0.06	
Calculation factor	e	0.8	
Calculation factor	X	0.63	
Calculation factor	Y_0	0.66	
Calculation factor	Y_1	0.78	
Calculation factor	Y_2	1.24	

Mass

11. EBE Faculty: Assessment of Ethics in Research Projects

Any person planning to undertake research in the Faculty of Engineering and the Built Environment at the University of Cape Town is required to complete this form before collecting or analysing data. When completed it should be submitted to the supervisor (where applicable) and from there to the Head of Department. If any of the questions below have been answered YES, and the applicant is NOT a fourth year student, the Head should forward this form for approval by the Faculty EIR committee: submit to Ms Zulpha Geyer (Zulpha.Geyer@uct.ac.za; Chem Eng Building, Ph 021 650 4791).Students must include a copy of the completed form with the final year project when it is submitted for examination. [89]

Name of Principal Researcher/Student: DARREN RICHARD DUDLEY **Department:** ELECTRICAL ENGINEERING

If a Student: YES **Degree:** MSc ELEC ENG **Supervisor:** AZEEM KHAN

If a Research Contract indicate source of funding/sponsorship: N/A

Research Project Title: DESIGN OF A VERNIER PERMANENT MAGNET WIND GENERATOR

Overview of ethics issues in your research project:

Question 1: Is there a possibility that your research could cause harm to a third party (i.e. a person not involved in your project)?	YES	X-NO
Question 2: Is your research making use of human subjects as sources of data? If your answer is YES, please complete Addendum 2.	YES	X-NO
Question 3: Does your research involve the participation of or provision of services to communities? If your answer is YES, please complete Addendum 3.	YES	X-NO
Question 4: If your research is sponsored, is there any potential for conflicts of interest? If your answer is YES, please complete Addendum 4.	YES	X-NO

If you have answered YES to any of the above questions, please append a copy of your research proposal, as well as any interview schedules or questionnaires (Addendum 1) and please complete further addenda as appropriate.

I hereby undertake to carry out my research in such a way that

- there is no apparent legal objection to the nature or the method of research; and
- the research will not compromise staff or students or the other responsibilities of the University;
- the stated objective will be achieved, and the findings will have a high degree of validity;
- limitations and alternative interpretations will be considered;
- the findings could be subject to peer review and publicly available; and
- I will comply with the conventions of copyright and avoid any practice that would constitute plagiarism.

Signed by:

	Full name and signature	Date
Principal Researcher/Student:	**Darren Richard Dudley**	13 October 2020

This application is approved by:

Supervisor (if applicable):	**Professor Azeem Khan**	13 October 2020
HOD (or delegated nominee): Final authority for all assessments with NO to all questions and for all undergraduate research.	**Janine Buxey**	13 October 2020
Chair : Faculty EIR Committee For applicants other than undergraduate students who have answered YES to any of the		

EBE Faculty: Assessment of Ethics in Research Projects

Any person planning to undertake research in the Faculty of Engineering and the Built Environment at the University of Cape Town is required to complete this form before collecting or analysing data. When completed it should be submitted to the supervisor (where applicable) and from there to the Head of Department. If any of the questions below have been answered YES, and the applicant is NOT a fourth year student, the Head should forward this form for approval by the Faculty EIR committee: submit to Ms Zulpha Geyer (Zulpha.Geyer@uct.ac.za; Chem Eng Building, Ph 021 650 4791). Students must include a copy of the completed form with the thesis when it is submitted for examination.

Name of Principal Researcher/Student: DARREN DUDLEY Department: ELECTRICAL ENGINEERING

If a Student: Degree: BSc Eng - Elec Eng Supervisor: PROF AZEEM KHAN

If a Research Contract indicate source of funding/sponsorship:

Research Project Title: Design of a Vernier Permanent Magnet Wind Generator

Overview of ethics issues in your research project:

Question 1: Is there a possibility that your research could cause harm to a third party (i.e. a person not involved in your project)?	YES	NO ✓
Question 2: Is your research making use of human subjects as sources of data? If your answer is YES, please complete Addendum 2.	YES	NO ✓
Question 3: Does your research involve the participation of or provision of services to communities? If your answer is YES, please complete Addendum 3.	YES	NO ✓
Question 4: If your research is sponsored, is there any potential for conflicts of interest? If your answer is YES, please complete Addendum 4.	YES	NO ✓

If you have answered YES to any of the above questions, please append a copy of your research proposal, as well as any interview schedules or questionnaires (Addendum 1) and please complete further addenda as appropriate.

I hereby undertake to carry out my research in such a way that

- there is no apparent legal objection to the nature or the method of research; and
- the research will not compromise staff or students or the other responsibilities of the University;
- the stated objective will be achieved, and the findings will have a high degree of validity;
- limitations and alternative interpretations will be considered;
- the findings could be subject to peer review and publicly available; and
- I will comply with the conventions of copyright and avoid any practice that would constitute plagiarism.

Signed by:

	Full name and signature	Date
Principal Researcher/Student: DARREN DUDLEY	DARREN RICHARD DUDLEY [signature]	17/04/15

This application is approved by:

Supervisor (if applicable):	Azeem Khan [signature]	17/04/15
HOD (or delegated nominee): Final authority for all assessments with NO to all questions and for all undergraduate research.	[signature]	[illegible]
Chair : Faculty EIR Committee For applicants other than undergraduate students who have answered YES to any of the above questions.		

www.ingramcontent.com/pod-product-compliance
Lightning Source LLC
LaVergne TN
LVHW091430190726
843491LV00006B/1677

* 9 7 8 3 3 8 4 2 2 7 9 1 1 *